SpringerBriefs in Earth Sciences

For further volumes:
http://www.springer.com/series/8897

Jérôme Weiss

Drift, Deformation, and Fracture of Sea Ice

A Perspective Across Scales

Foreword by Alan S. Thorndike

 Springer

Jérôme Weiss
Laboratoire de Glaciologie et
  Géophysique de l'Environnement
Saint-Martin d'Hères cedex
France

ISSN 2191-5369
ISBN 978-94-007-6201-5
DOI 10.1007/978-94-007-6202-2
Springer Dordrecht Heidelberg New York London

ISSN 2191-5377 (electronic)
ISBN 978-94-007-6202-2 (eBook)

Library of Congress Control Number: 2013931216

© The Author(s) 2013

This work is subject to copyright. All rights are reserved by the Publisher, whether the whole or part of the material is concerned, specifically the rights of translation, reprinting, reuse of illustrations, recitation, broadcasting, reproduction on microfilms or in any other physical way, and transmission or information storage and retrieval, electronic adaptation, computer software, or by similar or dissimilar methodology now known or hereafter developed. Exempted from this legal reservation are brief excerpts in connection with reviews or scholarly analysis or material supplied specifically for the purpose of being entered and executed on a computer system, for exclusive use by the purchaser of the work. Duplication of this publication or parts thereof is permitted only under the provisions of the Copyright Law of the Publisher's location, in its current version, and permission for use must always be obtained from Springer. Permissions for use may be obtained through RightsLink at the Copyright Clearance Center. Violations are liable to prosecution under the respective Copyright Law.
The use of general descriptive names, registered names, trademarks, service marks, etc. in this publication does not imply, even in the absence of a specific statement, that such names are exempt from the relevant protective laws and regulations and therefore free for general use.
While the advice and information in this book are believed to be true and accurate at the date of publication, neither the authors nor the editors nor the publisher can accept any legal responsibility for any errors or omissions that may be made. The publisher makes no warranty, express or implied, with respect to the material contained herein.

Photograph on p. ii "Sunset on the ice" taken by Lucas Girard

Printed on acid-free paper

Springer is part of Springer Science+Business Media (www.springer.com)

Foreword

In 1967 I was an undergraduate, studying a little of this and a little of that. One afternoon, my geology professor, Jelle de Boer, took several of us to Amherst College to hear J. Tuzzo Wilson talk about plate tectonics and seafloor spreading. Wilson referred to a type of fault found at mid-ocean ridges—the transform fault. To illustrate the idea, he showed an areal photograph of a field of sea ice. A zig-zag fracture ran through the image with shearing motion along the zigs and opening or closing along the zags. This was the first time I had ever heard of sea ice. I suppose it deserves only a minor footnote in the history of earth science that the deformation of sea ice was on the mind of one of the early architects of plate tectonics.

More footnotes come to mind, acknowledging the role sea ice has played in developing ideas in other fields. Ekman's study of the boundary layers in the atmosphere and ocean was motivated by Nansen's observation that sea ice drifts to the right of the wind. Ekman's analysis of inertial oscillations in geophysical fluids began with the observation that sea ice at times follows a loopy path. And in our own era, sea ice participates in one of the mechanisms for climate change. Indeed, the positive feedback between the ice extent and the global albedo leads one to think that the climate response to changes in atmospheric composition will be amplified in the polar regions. Coarse oceanic sediments, picked up by sea ice along the beaches of the Arctic Ocean and deposited in the North Atlantic, place constraints on climates of the past. Finally, in the study of turbulent motion in fluids and granular materials, the motion of sea ice serves as an interesting example. I say nothing about Europa.

The first measurements of sea ice motion were made by scientists at ice camps using techniques of celestial navigation. The data rate was low, and often there was only one ice camp in operation. The measurements of latitude and longitude were needed to provide geographical reference for the oceanographic and atmospheric data being gathered at the ice camps. As these data accumulated, it became clear that while there were patterns to the ice motion, the patterns were accompanied by variability in somewhat the same way as the wind at some point can be thought to have an average value plus a seasonally varying part, plus a part due to the current weather system, plus a part due to the whatever turbulent conditions there may be.

In the 1960s and 1970s a number of groups were developing theoretical models of sea ice, models that, when given the winds and currents and heat fluxes, and displacements of the boundaries, would predict the thickness, extent, and motion of the ice. These models were deterministic in the sense that for given inputs the models calculated definite values for the output ice velocities and thicknesses. In retrospect, it seems strange that more effort was not invested in statistical models, models whose outputs were means and variances, correlations, and covariances. These statistics are what makes sea ice. The granularity of its geometry and its piecewise rigid motion should find expression in these statistics.

This book presents a statistical description of sea ice motion. The fundamental concept of velocity as a limit is examined, honoring the lead of L. F. Richardson, who asked in a famous paper if the wind possessed a velocity. The data suggest that the ice indeed does possess a velocity. But lest you think the author splits hairs, the ice velocity is not differentiable in space, so that many of the notions of continuum mechanics cannot have their usual meanings. As a function of space, the velocity is that of a rigid body, with discontinuities at the boundaries between plates. These plates may be single floes or huge assemblages of floes moving as a rigid body.

The critical statistic is the correlation between the velocities at two points. One expects it to be high if the points are close together, and low if far apart. The situation is a little easier to think about if we consider the scalar surface pressure rather than the vector ice velocity. Years ago I calculated the cross-correlation between the atmospheric pressure at two points and found correlations near unity for points close together, and decreasing correlation with distance. In fact the correlation coefficient became negative for separations of about 1000 km. This seemed plausible: if a point was experiencing an unusually high pressure, there must be places some distance away where the pressure was unusually low. But what would happen for three points at the vertices of a 1000 km triangle? Could three time series, A, B, and C, be such that the correlation between any two, $r_{AB} = <AB>$, $r_{AC} = <AC>$, and $r_{BC} = <BC>$ was negative? Clearly, $r_{AB} = r_{AC} = r_{BC} = -1$ is not possible. But what about, $r_{AB} = r_{AC} = r_{BC} = -0.6$? It turns out that the constraint on the correlation coefficients is $r_{AB}^2 + r_{AC}^2 + r_{BC}^2 < 1 + 2\, r_{AB} \times r_{AC} \times r_{BC}$. Bells' inequalities in quantum mechanics are a special case of this general result. But you will never find a footnote in a quantum mechanics text acknowledging the role of sea ice in the development of that field.

Dr. Weiss explains how to use the time and space lag correlation functions to test what has previously been assumed: namely that there exist space and time-scales for which sea ice behaves like a continuum. This provides support for the many sea ice models developed over the last half century. A wise man, in an essay about citation strategies, advised authors never to reveal who gave them the original idea behind their work. Weiss ignores this advice when he refers to Benoit Mandlebrot's Fractal Geometry of Nature. Every page of this book contains ideas developed by Mandlebrot. I remember the excitement I felt when Mandlebrot's book appeared. For years, fractals were my hammer and every sea ice problem was my nail. By now, the ideas have matured. In Weiss's capable hands, they are used to characterize the motion of sea ice.

The sea ice described by Zubov was about the same as Nansen had seen. And what Untersteiner encountered in the international geophysical year 1957 was no different. Ice grew to about 3 m in thickness, it moved a few kilometers a day, and it was fractured into discrete floes. Deformation occurred at the floe boundaries. The overall motion consisted of a clockwise gyre in the Beaufort Sea and a mighty river of ice traversing the Arctic Ocean from the Siberian coast to the Fram Strait. It was easy to think of the sea ice environment as unchanging from year to year. We had only to understand the seasonal cycle. Weiss's ice is more subtle. It varies on all space and time scales. We have developed models that do a fair job reproducing the seasonal and shorter time scales, only to find that they are worthless as forecasting tools. As evidence accumulates around us, we begin to see the arctic as a more dynamic place. We have dressed up Nansen's rule so that it accounts for the ice drift, but a greater challenge faces us. It is to understand the role of sea ice in a changing climate. At least part of that challenge is to understand what properties of the ice or the wind or ocean cause the ice motion to have the statistical structure that Weiss describes.

Oxford, MD, 26 November 2012   Alan S. Thorndike

Acknowledgments

In the course of his research on sea ice, the author benefited from the work and the expertise of many colleagues. First of all, he would like to thank his former Master or Ph.D. students, Pierre Rampal, Lucas Girard, Florent Gimbert, and Sebastien Marcq, whose research results are detailed in many places in this book. He would like also to warmly acknowledge (in alphabetical order): David Amitrano, Bernard Barnier, Sylvain Bouillon, Mickaël Bourgoin, Daniel Feltham, Thierry Fichefet, Jean-Claude Gascard, Jari Haapala, Nicolas Jourdain, Ron Lindsay, David Marsan, Jean-Philippe Metaxian, Erland Schulson, and Harry Stern. The home institution of the author, CNRS, is greatly acknowledged for allowing him to freely explore new research directions. Finally, Harry Stern and Pierre Rampal are warmly thanked for their detailed review of this book which greatly helped to improve its quality.

Further Readings

This small book deals with sea ice drift, deformation, and fracturing, with a special focus on scaling properties. For a more general description of ice (including sea ice) mechanics, see *Creep and Fracture of Ice* by Schulson and Duval (2009). Leppäranta (2005) proposed a thorough analysis of sea ice drift and momentum balance. Finally, the classical *Geophysics of Sea Ice* by Untersteiner (1986) remains nowadays, 25 years after its publishing, a great source of inspiration for the sea ice scientist, with, for the author of the present book, a special mention of Chap. 7 on *Kinematics of Sea Ice* by A.S. Thorndike (1986).

References

Leppäranta, M. (2005). *The drift of sea ice*. Berlin: Springer.
Schulson, E. M., & Duval, P. (2009). *Creep and fracture of ice*. Cambridge: Cambridge University Press.
Thorndike, A. S. (1986). Kinematics of sea ice. In N. Untersteiner (Ed.), *The geophysics of sea ice* (pp. 489–549). New York: Plenum Press.
Untersteiner, N. (1986). *The geophysics of sea ice*. New York: Plenum Press.

Contents

1	**Introduction**	1
	1.1 Sea Ice Kinematics: From the Fram's Journey to Thorndike's Legacy	1
	1.2 The Momentum Equation of Sea Ice	4
	1.3 Scaling: Some Basic Definitions	6
	References	8
2	**Sea Ice Drift**	11
	2.1 Data	13
	2.2 How to Extract a Mean Field	14
	2.3 Diffusion Regimes	18
	2.4 Turbulent-Like Fluctuations	21
	2.5 Sea Ice Acceleration and the Dynamical Origin of Intermittency	24
	2.6 Spectral Analysis	26
	2.7 Concluding Remarks	27
	References	28
3	**Sea Ice Deformation**	31
	3.1 Data	32
	3.2 Spatial Scaling and Localization of Deformation	34
	3.3 Spatial and Temporal Scaling Laws from the Dispersion of Lagrangian Trajectories	40
	3.4 Space/Time Coupling	44
	3.5 Sea Ice Dispersion as the Result of "Solid Turbulence"	46
	3.6 Spectral Analysis	47
	3.7 Concluding Remarks	48
	References	49
4	**Sea Ice Fracturing**	53
	4.1 Data	54
	4.2 Sea Ice Internal Stresses, Strength, and Rheology	56
	4.3 Intermittency of Sea Ice Stresses	60

	4.4 Fracture Networks	62
	4.5 A Statistical Model of Sea Ice Fracturing and Deformation	65
	References	69
5	**Conclusion and Perspectives: Sea Ice Drift, Deformation and Fracturing in a Changing Arctic**	73
	References	81

Symbols

c	Strength of the space/time coupling (Sect. 3.4)
$C(t)$	Autocorrelation function
C_a	Air drag coefficient
C_w	Water drag coefficient
D	Fractal dimension
$f = 2\Omega \sin \phi$	Coriolis parameter
G	Geostrophic wind
h	Ice thickness
H	"Roughness" exponent
k	Vertically upward unit vector
l	Defect, or "stress concentrator" size (Sect. 4.2)
L	Length scale (generic)
L_η	Kolmogorov dissipation length scale of turbulence
$\mathcal{M}$	Measure (generic): Sect. 1.3
$M \sim whl$	Seismic moment
N_a	Nansen's number
p_L	Fraction of open water (fracture density) measured at scale L (Sect. 4.3)
q	Order of a moment
R_e	Reynolds number
Δr	Change in separation of pairs of tracers
s	Scale (generic): Sect. 1.3 Floe size (or diameter): Chap. 4
t	Time
t_η	Kolmogorov dissipation timescale of turbulence
U = (u,v)	Velocity vector
U$'$	Velocity fluctuation (velocity—mean flow)
U	Speed (norm of the velocity)
U'	Speed fluctuation
U$_\mathbf{a}$	Wind velocity vector
U$_\mathbf{i}$	Sea ice velocity vector
U_i	Sea ice speed (norm of the velocity)
U'_i	Sea ice speed fluctuation
U$_\mathbf{w}$	Water velocity vector

w	Lead width: Sect. 4.3 Offset of a fracturing/faulting event, averaged over its length l
$\mathbf{X} = (x,y)$	Position vector
$\mathcal{X}, \mathcal{Y}$	Random variables (generic)
$\alpha, \beta, \chi, \eta, \gamma, \zeta, \varsigma$	Scaling exponents
δ	Fractal exponent in time: Sect. 4.4
ε	Strain (generic)
$\dot{\varepsilon}$	Strain-rate (generic)
$\dot{\varepsilon}_{shear}$	Shear strain-rate
$\dot{\varepsilon}_{div}$	Divergence rate
$\dot{\varepsilon}_{tot}$	Total strain-rate (combination of shear and divergence)
$\dot{\varepsilon}_{disp}$	Strain-rate proxy obtained from the dispersion of pairs of trajectories
$\dot{\varepsilon}_{D/B}$	Ductile-to-brittle transition strain-rate
ϕ	$\text{Arctan}\left(\frac{\dot{\varepsilon}_{shear}}{\dot{\varepsilon}_{div}}\right)$, invariant indicating the mode of deformation
φ	Intersection angle between shear faults of a conjugate set (related to the internal friction coefficient μ_i, see Sect. 4.3)
Γ	Integral time
μ	Friction coefficient
μ_i	Internal friction coefficient
θ	Turning angle (Chaps. 1 and 2) Principal stress direction (Sect. 4.2)
ρ_a	Air density
ρ_i	Sea ice density
ρ_w	Water density
σ	Internal ice stress
σ_I	First principal stress (in the horizontal plane of the ice cover)
σ_{II}	Second principal stress (in the horizontal plane of the ice cover)
σ_N	Normal stress, or "pressure", $\sigma_N = (\sigma_I + \sigma_{II})/2$
σ_c	Compressive strength
τ	Drag force per unit ice surface (generic): Sect. 1.2 Timescale: Sects. 2.2 and 2.3 Shear stress $\tau = (\sigma_I - \sigma_{II})/2$: Chap. 4
τ_0	Cohesion
τ_c	Shear strength
τ_a	Wind "stress" (drag force per unit ice surface)
τ_w	Water "stress" (drag force per unit ice surface)
Ω	Earth's angular velocity
ω	Frequency

Chapter 1
Introduction

Abstract In this introductory chapter, after a short historical perspective on sea ice drift studies from the pioneering observations of Nansen, we briefly introduce the momentum equation of sea ice as well as some generic definitions, tools and concepts that will be used afterwards in this book to analyze the scaling properties of the sea ice cover.

1.1 Sea Ice Kinematics: From the Fram's Journey to Thorndike's Legacy

Sea ice drift, deformation, and fracturing have been known empirically for thousands of years by the indigenous people of the Arctic. Traveling over sea ice to hunt seals or to fish, they knew the danger of a sudden breakup of the ice cover, but also used drifting ice floes as hunting stations in spring. In coastal regions where ice remains generally "fastened" to the shore, the so-called fast ice does not drift or deform, remaining essentially smooth and providing a convenient way to travel on sledges during winter.

However, if fast ice facilitates transportation along the coasts, sea ice represents a formidable obstacle to navigation. Consequently, the Arctic Ocean, which was essentially covered by ice year-round until recently, remained Oceana Incognita until the end of the nineteenth century. For a long time, navigators and explorers dreamed of a northern passage from Europe to Asia as an alternative to the spice trade routes. At the end of the sixteenth century, Willem Barents navigated to Novaya Zemlya three times, but became stuck in sea ice in July 1596, and died after a winter spent onshore. Three centuries later, Adolf Erik Nordenskiöld, onboard the *Vega*, was the first to navigate the Northern Sea Route in 1878. At almost the same time, in July 1879, unaware of the success of Nordenskiöld, George Washington De Long, captain of the *Jeannette*, left San Francisco for the Bering Strait, in order to try the Northern Sea Route from the east. The ship became stuck in sea ice in September 1879, and drifted erratically with the ice cover for almost two years (Fig. 1.1). In June 1881, north of the New Siberian Islands, compressive ice forces crushed the ship's hull, and the *Jeannette* sank into the Arctic Ocean.

Surprisingly, fragments of the *Jeannette* were found three years later on the south coast of Greenland. This led Fridtjof Nansen to conceive a visionary voyage across the Arctic Ocean, aided by a hypothetical current, the Transpolar Drift. Nansen's ship, the *Fram*, was specially designed to resist compressive ice forces and to ride up over the ice cover in case of convergent deformation. The expedition started in June 1893 with a crew of 12, led by Otto Sverdrup. Following the Northern Sea Route, the *Fram* rounded Cape Laptev and reached the Laptev Sea in September, where she got stuck in the ice at about $78°30'N$. During the first few months the *Fram* drifted slowly northward, but in a very erratic way, zigzagging in different directions—a first illustration of the extreme variability and intermittency of sea ice drift, a point discussed in detail in Chap. 2. During the expedition, Nansen, Sverdrup and the crew performed numerous meteorological, oceanographic and sea ice measurements (Nansen 1902). Their position was determined every two days from astronomical observations, giving the first-ever dataset on sea ice drift in relation to wind forcing and ocean currents. These measurements showed that sea ice drifts on average at 2 % of the wind speed, and with a deviation of 20–40° to the right of the wind direction (Nansen 1902). This last result led Ekman (1905) to develop his theory of inertial oscillations and the role of Earth's rotation on ocean currents. Sea ice inertial oscillations will be briefly discussed in Chap. 2. After three years of an incredible journey, the *Fram* passed through what is now called Fram Strait and reached the open North Atlantic Ocean, with the entire crew safe.

Oceanographic measurements performed during the Fram's drift revealed a deep Arctic Ocean with stratification: a cold, relatively fresh surface layer of 50–100 m overlying warm (>0 °C) and salty intermediate water (~ 150–900 m). Nansen (1902) had already suggested that the warm layer came from the North Atlantic, entering the Arctic Ocean through Fram Strait and the Barents Sea, a hypothesis confirmed in modern times (e.g. Polyakov et al. 2010).

After this pioneering work, observations and analyses of sea ice motion and mechanics remained scarce during the first half of the twentieth century, mainly restricted to Soviet drifting stations and camps (Zubo 1945). A further important step was taken in the 1970s with the Arctic Ice Dynamics Joint Experiment (AIDJEX), which undertook field measurements in the Beaufort Sea and led to progress in the characterization and modeling of sea ice kinematics, dynamics, and mechanics (Untersteiner 1980). In particular, a set of ice drifters with an automatic ARGOS positioning system was launched. This set of data was used by A.S. Thorndike and co-workers to develop the first stochastic analysis of ice motion (Thorndike 1986b). The relationship between sea ice motion and wind forcing was analyzed in Thorndike and Colony (1982), whereas Colony and Thorndike (1984) decomposed the ice drift into a mean field and a random walk. Thorndike also used the statistical tools of turbulence to analyze sea ice drift

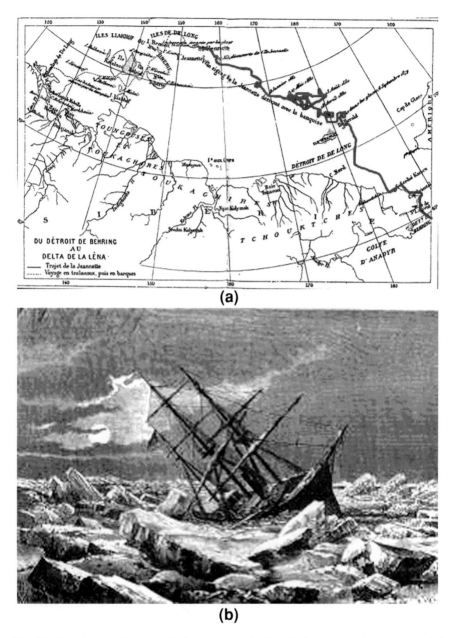

Fig. 1.1 The dramatic journey of the *Jeannette* in the Arctic: one of the first recorded Lagrangian trajectories of sea ice. **a** A map showing the tortuous trajectory of the ship in the East Siberian Sea. **b** The *Jeannette* crushed by compressive ice forces. (From *L'expédition de la Jeannette au pôle nord racontée par tous les membres de l'expédition*, M. Dreyfous ed., Paris, 1883)

(Thorndike 1986a) and deformation (Martin and Thorndike 1985). We will come back to these issues throughout this volume.

Since AIDJEX, the amount of data available has increased considerably: ice drifters (buoys) have been launched every year in the Arctic by the International Arctic Buoy Program (IABP), and satellite data, particularly synthetic aperture radar (SAR) imagery, has furnished an unprecedented dataset on sea ice motion and deformation fields over the Arctic (Fily and Rothrock 1987; Kwok et al. 1990). These data represent an unrivaled source of information on the deformation and fracturing of a geophysical material across a wide range of spatial (1–1000 km) and temporal (1 h–30 years) scales. Taking advantage of these new data and inspired by Thorndike's legacy, new analyses were performed in the last decade to better characterize and understand the physics of sea ice drift, deformation and fracturing. At the same time, three decades of uninterrupted observations revealed spectacular changes in Arctic sea ice kinematics and mechanics, associated with the thinning and decline of the sea ice cover in the context of climate change. All these recent developments are the subject of the present volume.

For various reasons, most of the observations and analyses on sea ice kinematics have been performed in the North (Arctic ocean, Baltic sea,...), although there are recent developments in the South (e.g. Hutchings et al. 2012). Consequently, most of the results presented below have been obtained for the Arctic sea ice cover. Although Antarctic sea ice differs from Arctic sea ice in terms of annual cycle (perennial Antarctic sea ice is limited in extent) and boundary conditions (an ocean surrounded by continents in the Arctic, the reverse in the Antarctic), we believe that the fundamental physical mechanisms discussed in this volume are relevant for both hemispheres.

1.2 The Momentum Equation of Sea Ice

The sea ice cover drifts, deforms, and fractures under the action of external forces, essentially wind and ocean currents, and subject to boundary conditions (open boundary, coastline geometry). Owing to the very small aspect ratio between its average thickness ($\sim$m) and its lateral extent ($>10^3$ km for the Arctic basin), we consider the ice cover to be a 2D object and neglect motions and forces out of the sea surface plane. Assuming further a continuum mechanics description—a questionable hypothesis as we will show later—the general form of the equation of motion of sea ice can be written as:

$$\rho_i h \left[\frac{\partial \mathbf{U}_i}{\partial t} + \mathbf{U}_i \cdot \nabla \mathbf{U}_i + f \mathbf{k} \times \mathbf{U}_i \right] = \nabla \cdot \sigma + \tau_a + \tau_w \quad (1.1)$$

where ρ_i is ice density, h the ice thickness, $\mathbf{U}_i = (u, v)$ the (horizontal) ice velocity, $f = 2\Omega \sin \phi$ the Coriolis parameter with ϕ the latitude and $\Omega = 7.292 \times 10^{-5}\ \mathrm{s}^{-1}$ the Earth's angular velocity, σ the internal ice stress (integrated over the thickness,

1.2 The Momentum Equation of Sea Ice

i.e. in N·m^{-1}) and τ_a and τ_w respectively the air and water "stresses" (frictional forces per m^2 of ice surface). On the left-hand side of Eq. (1.1), $\frac{\partial U_i}{\partial t}$ is a local acceleration, whereas $U_i \cdot \nabla U_i$ is an advective acceleration (Leppäranta 2005). The last term of the left hand side can be moved to the right hand side as a Coriolis "force". Other terms such as the sea surface tilt or a term related to the air pressure gradient can be added to Eq. (1.1), but are generally considered as negligible (Steele et al. 1997). For more details about the derivation of the momentum equation of sea ice, see e.g. Leppäranta (2005).

The horizontal divergence $\nabla \cdot \sigma$ represents the force due to the internal stress field, which is assumed to encompass different kinds of solid mechanical interactions within the ice cover such as frictional contact forces between ice floes, shearing along faults/leads, and crushing during convergent deformation and ridge formation. In this continuum mechanics formulation, all these processes are summarized in a single force applied to the ice parcel, and a rheology, i.e. a relationship between stresses and strains, is needed to calculate σ. Usually, wind stress (τ_a) is considered to be the main driving force of sea ice motion, balanced by ice-ocean drag (τ_w) and the "internal friction" $\nabla \cdot \sigma$ (Steele et al. 1997) (see Fig. 1.2). The Coriolis force is smaller than these three terms in magnitude and always perpendicular to the ice motion.

For a turbulent flow, classical dimensional analysis gives a square dependence of the drag force on the fluid speed, $\tau = \rho C U^2$, where ρ is the fluid density, U the fluid speed, and C a dimensionless drag coefficient. For laminar flow, i.e. low Reynolds number R_e, $C \sim R_e$ and consequently $\tau \sim U$. Although a linear drag

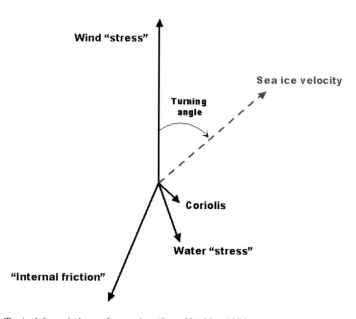

Fig. 1.2 Typical force balance for sea ice (from Hunkins 1975)

law can be a reasonable approximation in case of water drag (Gimbert et al. 2012b; Heil and Hibler 2002), a quadratic form is generally preferred for both water and air drag (e.g. McBean 1986; McPhee 1986):

$$\tau_a = \rho_a C_a |\mathbf{U}_a| \mathbf{U}_a \tag{1.2a}$$

$$\tau_w = \rho_w C_w |\mathbf{U}_w - \mathbf{U}_i| (\mathbf{U}_w - \mathbf{U}_i) \tag{1.2b}$$

where indices a and w stand respectively for air and water. In these equations, surface layer velocities are considered, and ice velocity is omitted in Eq. (1.2a) as $|U_a| \gg |U_i|$. If geostrophic velocities are considered instead, constant turning angles can be added to these expressions.

Each of the force terms on the right-hand side of Eq. (1.1) is a source of complexity for sea ice motion. Indeed, the main driving force, wind, is turbulent by nature, i.e. is characterized by fluctuations over a large range of spatial and temporal scales with associated scaling laws (e.g. Frisch 1995). As discussed in more detail below (Chap. 4), sea ice behaves as an elasto-brittle medium, i.e. its response to mechanical forcing is strongly non-linear and involves long-range elastic interactions. These two ingredients are a source of spatial heterogeneity, intermittency and complex mechanical interactions over a broad range of scales, as is observed e.g. for the Earth's crust. The combination of a complex forcing and non-linear response makes the analysis of sea ice fracturing, deformation, and drift an interesting challenge. In particular, one underlying question is: what is inherited from the complexity of the forcing, or what results from the mechanical behavior of sea ice itself? As stressed throughout this volume, statistical and scaling analysis is the preferred tool to try to answer these questions and challenges.

1.3 Scaling: Some Basic Definitions

In a very broad sense, *scaling* designates the way an observable $\mathcal{M}$ changes with the scale of observation s. The scale s can be a spatial scale, a time scale, an energy scale, etc.

However, following the terminology used in statistical physics and especially in the physics of critical phenomena, which is used to establish links between very different scales, scaling can have a more restrictive meaning (see e.g. Stanley 1999). An observable $\mathcal{M}$ exhibits *scaling*, *scaling properties*, or *scale invariance* (all these terms are used in the literature) with respect to the scale s if:

$$\mathcal{M}(\lambda s) = \lambda^\alpha \mathcal{M}(s) \tag{1.3}$$

where λ is a scale factor and α the exponent associated with the function $\mathcal{M}(s)$. In this volume, scaling will be used with this restrictive meaning. The only function that satisfies (1.3) is a power law (in what follows, the sign "$\sim$" is used to indicate proportionality):

1.3 Scaling: Some Basic Definitions

$$\mathcal{M}(s) \sim s^{\alpha} \tag{1.4}$$

This explains why *power law behavior* is another term for scaling. Equation (1.3), which is a *scaling law*, can be extended to multiple scales $s_1, s_2,...$, with possibly different exponents associated with the scales. This can be expressed as:

$$\mathcal{M}(\lambda^{1/\alpha} s_1, \lambda^{1/\beta} s_2, \ldots) = \lambda \mathcal{M}(s_1, s_2, \ldots) \tag{1.5}$$

If two observables $\mathcal{M}_1$ and $\mathcal{M}_2$ show scaling properties with respect to the scale s, i.e. $\mathcal{M}_1(s) \sim s^{\alpha_1}$ and $\mathcal{M}_2(s) \sim s^{\alpha_2}$, then one has $\mathcal{M}_2(s) \sim (\mathcal{M}_1(s))^{\frac{\alpha_2}{\alpha_1}}$ which can be considered another expression of scaling.

Different examples of scaling laws associated with sea ice kinematics, deformation and fracturing will be detailed below. However, some general examples can be briefly presented here as illustrations.

If the scale s is a spatial scale and the observable is the box-counting measure of a geometrical set (the number N of boxes of size s needed to cover the set), scale invariance is characterized by:

$$N(\lambda s) = \lambda^{-D} N(s) \tag{1.6}$$

or,

$$N(s) \sim s^{-D} \tag{1.7}$$

Geometrical sets following such spatial scale invariance, with D different from an Euclidian dimension, are called *fractal* sets and D is a fractal dimension. See e.g. Mandelbrot (1982) or Feder (1988) for overviews on fractals.

Another example of scaling with s as a spatial scale is the expression for the number of objects of size s within a set, which can follow power law behavior, $N(s) \sim s^{-\alpha}$. Here, the notation $N(s)$ refers to a frequency distribution, i.e. to the number of objects between s and $s + ds$. The corresponding cumulative distribution, noted $N(>s)$, scales as $s^{-\alpha+1}$ if $\alpha > 1$. Let us consider the following practical example: it will be shown in Chap. 4 that sea ice, when intensively fractured, is fragmented into ice floes with a power law distribution of sizes, $N(s) \sim s^{-\alpha}$ with $\alpha \approx 3$. This means that the number of floes of size s, $N(s)$, is on average eight times larger than the number of floes of size $2s$, $N(2s)$, whatever s. In other words, looking at an aerial or satellite photograph of the sea ice cover, it is impossible to determine the scale of the photograph by comparing the relative number of floes of different sizes. The photographs "look the same" whatever their resolution.

Throughout this volume, the scaling properties of a field of values (e.g. a strain-rate field) over a spatial domain (e.g. the Arctic basin), or a time series over a time interval, will be analyzed. In these cases, a simple characterization by a fractal dimension such as in relations (1.6) and (1.7) is inadequate. Let us consider for example a strain-rate field of the sea ice over the Arctic basin in winter (see Fig. 3.1).

In theory, the strain-rate can be measured everywhere over the domain, i.e. the geometrical support of the measure has a topological dimension of 2. We want instead to characterize how the spatial distribution of the measure $\mathcal{M}(s)$ evolves with the scale of observation s. One way to do this is to determine the scaling properties of the probability density function (PDF) of the measure, $P(\mathcal{M}(s))$. Technically, to determine directly the scaling symmetries of the PDF is difficult. This is why an analysis of the moments $\langle \mathcal{M}(s)^q \rangle$ of order q of the distribution is preferred. Scaling of the field is then characterized by:

$$\langle \mathcal{M}(s)^q \rangle \sim s^{\beta(q)} \tag{1.8}$$

which defines a full set of exponents $\beta(q)$. If $\beta(q)$ is a non-linear function of q, the measure is said to be multifractal. Indeed, the renormalized moments $\langle \mathcal{M}(s)^q \rangle^{1/q}$, although all expressed in the units of $\mathcal{M}$, scale differently with s, as $\beta(q)/q$ is not constant. For large (respectively small) values of q, the renormalized moments of order q "sample" preferentially the regions where $\mathcal{M}$ is large (respectively small). Multifractal scaling implies that the regions of large or small $\mathcal{M}$ are characterized by different fractal generalized dimensions D_q (Hentschel and Procaccia 1983) obtained from the exponents $\beta(q)$ and the topological dimension of the support of the measure. The full set of dimensions (or of exponents $\beta(q)$) fully characterizes the heterogeneity and clustering properties of the field. Another interpretation is that a multifractal field is fractal in two ways: (i) it is characterized by local singularities of the measure associated with an exponent $\alpha(\mathbf{X})$ that depends on position $\mathbf{X}$, and (ii) the ensemble of positions where the local exponent takes a value α is itself a fractal structure of dimension $f(\alpha)$, defining a multifractal spectrum $f(\alpha)$ which can be related to the generalized dimensions D_q through a Legendre transform (Chhabra and Jensen 1989). Considering time series, a multifractal analysis characterizes the intermittency. Multifractal concepts were initially developed in the context of chaotic systems (Hentschel and Procaccia 1983) and turbulence (Chhabra et al. 1989; Frisch 1995).

References

Chhabra, A. B., Meneveau, C., Jensen, R. V., & Sreenivasan, K. R. (1989). Direct determination of the F (Alpha) singularity spectrum and its application to fully-developed turbulence. *Physical Review A, 40*(9), 5284–5294.

Colony, R., & Thorndike, A. S. (1984). An estimate of the mean field of the arctic sea ice motion. *Journal of Geophysical Research, 89*(C6), 10623–10629.

Ekman, V. W. (1905). On the influence of the Earth's rotation on ocean currents. *Arkiv for Mathematik, Astronomi och Fysik, 2*(11), 1–52.

Feder, J. (1988). *Fractals*. New York: Plenum Press.

Fily, M., & Rothrock, D. A. (1987). Sea ice tracking by nested correlations. *IEEE Transactions on Geoscience Remote Sensing, 25*(5), 570–580.

Frisch, U. (1995). *Turbulence. The legacy of A.N. Kolmogorov*. Cambridge: Cambridge University Press.

References

Gimbert, F., Jourdain, N. C., Marsan, D., Weiss, J., & Barnier B. (2012b). Recent mechanical weakening of the Arctic sea ice cover as revealed from larger inertial oscillations. *Journal of Geophysical Research, 117*, C00J12.

Heil, P., & Hibler, W. D. I. (2002). Modeling the high frequency component of arctic sea ice drift and deformation. *Journal of Physical Oceanography, 32*, 3039–3057.

Hentschel, H. G. E., & Procaccia, I. (1983). The infinite number of generalized dimensions of fractals and strange attractors. *Physica D, 8*, 435–444.

Hunkins, K. (1975). Oceanic boundary-layer and stress beneath a drifting ice floe. *Journal of Geophysical Research-Oceans and Atmospheres, 80*(24), 3425–3433.

Hutchings, J. K., Heil, P., Steer, A., & Hibler, W. D. (2012). Subsynoptic scale spatial variability of sea ice deformation in the western Weddell Sea during early summer. *Journal of Geophysical Research-Oceans, 117*, C01002.

Kwok, R., Curlander, J. C., McConnell, R., & Pang, S. S. (1990). An ice-motion tracking system at the Alaska SAR facility. *IEEE Journal of Oceanic Engineering, 15*(1), 44–54.

Leppäranta, M. (2005). *The drift of sea ice*. Berlin: Springer.

Mandelbrot, B. B. (1982). *The fractal geometry of nature*. New York: W.H. Freeman.

Martin, S., & Thorndike, A. S. (1985). Dispersion of sea ice in the Bering Sea. *Journal of Geophysical Research, 90*(C4), 7223–7226.

McBean, G. (1986). The atmospheric boundary layer. In N. Untersteiner (Ed.), *The Geophysics of Sea Ice* (pp. 283–338). New York: Plenum Press.

McPhee, M. G. (1986). The upper ocean. In N. Untersteiner (Ed.), *The geophysics of sea ice* (pp. 339–395). New York: Plenum Press.

Nansen, F. (1902). Oceanography of the north polar basin: The Norwegian north polar expedition 1893–96, *Scientific Results, 3*(9), 427.

Polyakov, I. V., et al. (2010). Arctic Ocean warming contributes to reduced polar ice cap. *Journal of Physical Oceanography, 40*(12), 2743–2756.

Stanley, H. E. (1999). Scaling, universality, and renormalization: The three pillars of modern critical phenomena. *Reviews of Modern Physics, 71*(2), S358–S366.

Steele, M., Zhang, J., Rothrock, D., & Stern, H. (1997). The force balance of sea ice in a numerical model of the Arctic Ocean. *Journal of Geophysical Research, 102*(C9), 21061–21079.

Thorndike, A. S. (1986a). Diffusion of sea ice. *Journal of Geophysical Research, 91*(C6), 7691–7696.

Thorndike, A. S. (1986b). Kinematics of sea ice. In N. Untersteiner (Ed.), *The geophysics of sea ice* (pp. 489–549). New York: Plenum Press.

Thorndike, A. S., & Colony, R. (1982). Sea ice motion in response to geostrophic winds. *Journal of Geophysical Research, 87*(C8), 5845–5852.

Untersteiner, N. (1980). AIDJEX Review. In R. Pritchard (Ed.), *Sea ice processes and models*. Seattle: University of Washington Press.

Zubov, N. N. (1945). Arctic Ice, English translation by the AMS, San Diego, 1963, Moscow.

Chapter 2
Sea Ice Drift

Abstract In this chapter, sea ice drift is discussed in terms of statistical properties of individual Lagrangian trajectories. It is first shown that the Arctic sea ice velocity field can be objectively decomposed into a mean field and fluctuations. The mean field shows intra-annual (between winter and summer) as well as some interannual variability. The fluctuations, defined as the remaining part of the velocity field after removing the mean field, share similarities with fluid turbulence, such as a Kolmogorov-like scaling of the power spectral density, a ballistic regime within an inertial range of motion, or intermittency. However, significant differences are also observed: sea ice velocity distributions are clearly much more spread than Gaussian statistics, intermittency is more pronounced, and sea ice accelerations cannot be explained from wind stress statistics. These differences argue for a non-linear response of sea ice to forcing.

As mentioned in Chap. 1, (Nansen 1902) reported that sea ice drifts an average of about 2 % of the wind speed, and 20–40° to the right of the wind direction. A first order justification of this 2 % rule of thumb can be obtained from the momentum equation (1.1) with several (rough) simplifying assumptions. We consider a one-dimensional model of the free drift of sea ice, i.e. rotational effects are not considered and the internal stress term $\nabla \cdot \sigma$ is neglected. We consider further that the ocean is at rest ($\mathbf{U_w} = 0$) and the wind forcing $\mathbf{U_a}$ constant. Then the equation of motion reads (Leppäranta 2005):

$$\rho_i h \frac{d\mathbf{U_i}}{dt} = \rho_a C_a |\mathbf{U_a}|\mathbf{U_a} - \rho_w C_w |\mathbf{U_i}|\mathbf{U_i} \tag{2.1}$$

Taking as an initial condition $\mathbf{U_i}(t = 0) = 0$, the solution of Eq. (2.1) is:

$$\mathbf{U_i}(t) = \sqrt{\frac{\rho_a C_a}{\rho_w C_w}} \, \mathbf{U_a} \, \tanh\left(\sqrt{\frac{\rho_a C_a \rho_w C_w}{\rho_i h}} \mathbf{U_a} t\right) \tag{2.2}$$

J. Weiss, *Drift, Deformation, and Fracture of Sea Ice*,
SpringerBriefs in Earth Sciences, DOI: 10.1007/978-94-007-6202-2_2,
© The Author(s) 2013

which gives a steady-state solution $\mathbf{U}_i = Na\ \mathbf{U}_a$, where $Na = \sqrt{\frac{\rho_a C_a}{\rho_w C_w}}$ is the so-called Nansen number. Taking $\rho_a = 1.3$ kg/m^3, $\rho_w = 1025$ kg/m^3 for sea water, $C_a \approx 1.5 \times 10^{-3}$ and $C_w \approx 5 \times 10^{-3}$ (Leppäranta 2005), we obtain $Na = 0.0195$, in excellent agreement with Nansen's rule of thumb.

However, owing to the simplifying assumptions, such derivation can only give a very approximate view of sea ice drift, when averaged over large spatial and temporal scales and for conditions associated with low internal stresses, such as a loose summer ice cover. Using data from the first automatic ice drifters (in 1979 and 1980), Thorndike and Colony (1982) analyzed in more detail the response of sea ice to geostrophic wind forcing, i.e. the wind that results theoretically from the balance between the Coriolis effect and the pressure gradient. More specifically, they considered the following linear relationship:

$$\mathbf{U} = \mathbf{A}\mathbf{G} + \mathbf{C} + \mathbf{E} \qquad (2.3)$$

where $\mathbf{A} = Ae^{-i\theta}$ is a complex constant with θ the turning angle, $\mathbf{G}$ the geostrophic wind, $\mathbf{C}$ a constant mean velocity field assumed to be related to a mean ocean current, and $\mathbf{E}$ the residuals. As shown above, such a linear equation can be justified by a steady-state solution of the ice momentum equation in case of free-drift. Therefore, it ignores inertial effects, ice internal stresses, and non-geostrophic wind turbulence.

Using daily averages of sea ice velocity and surface pressure obtained from this limited dataset, Thorndike and Colony (1982) showed that typical values of the ratio $|A|$ and the turning angle θ are smaller than Nansen's estimate and vary with the season and the region of the Arctic, both parameters decreasing with increasing ice thickness and compactness, such as in winter and north of Greenland and the Canadian archipelago (from $A \approx 0.8\ \%$ and $\theta \approx 5°$ in winter, to $A \approx 1\ \%$ and $\theta \approx 20°$ in summer). These results, which can be qualitatively explained by the damping effect of internal stresses, were confirmed by Thomas (1999) from a larger dataset covering 15 years (1979–1993).

On the other hand, only 65–70 % of the variance of the ice velocity in the central Arctic is explained by the geostrophic wind, and even less within 400 km of the coasts or in peripheral seas (Thomas 1999; Thorndike and Colony 1982). The remaining variance could come from measurement error, small time-scale processes (including inertia), ocean current variability, or from the mechanical behavior of sea ice. Therefore, at this stage, several questions arise:

(i) Equation 2.3 considers a mean velocity field, which is generally calculated from simple averaging of velocity data at some arbitrary spatial scale (large enough to ensure a significant number of observations over each grid cell) and over the duration of the entire dataset. Are these averaging scales appropriate? Which physical processes are embodied in this mean circulation?

(ii) What is the physical origin of the remaining variance? Can we see the fingerprint of oceanic or atmospheric turbulence, or of ice internal stresses, in these fluctuations?

As shown below, proper statistical and scaling analyses are needed to answer these questions.

2.1 Data

There are basically two sources of sea ice drift data: Lagrangian trajectories of buoys and drifting stations, and satellite-derived velocities. Both have advantages and disadvantages.

In the early days of Arctic exploration, positions of drifting stations or ships were obtained from astronomical measurements (star shots during nighttime, sun shots during daylight), with daily temporal sampling at best and an accuracy of ~ 1 km (Nansen 1902; Reed and Campbell 1962). Since the end of the 1970s, automatic ice drifters (buoys) have been launched every year within the Arctic basin, and more recently and irregularly around the Antarctic (e.g. Hutchings et al. 2012). In the Arctic, the corresponding database is maintained by the International Arctic Buoy Program (IABP), see Fig. 2.1. Until recently, the positioning system of the buoys was ARGOS, based on a Doppler shift effect, with an accuracy of 300–500 m (Thomas 1999; Thorndike 1986b). The raw IABP trajectories are then smoothed by cubic interpolation, a procedure that likely reduces the positioning standard error to 100–150 m (Lindsay and Stern 2003). The ice velocity $\mathbf{U}$ is then calculated from successive positions $\mathbf{X}$, $\mathbf{U}(\mathbf{X}, t) = \frac{\mathbf{X}(t+\tau) - \mathbf{X}(t)}{\tau}$, where τ is here the temporal sampling of the measure (the dependence of the estimate of $\mathbf{U}$ on time

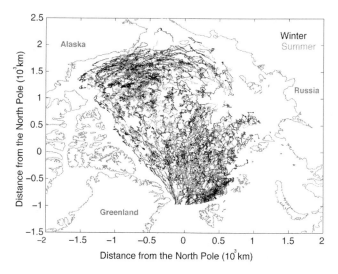

Fig. 2.1 IABP buoys trajectories recorded between December 1978 and December 2001 in the Arctic basin. The tortuous character of sea ice motion is obvious (adapted from Rampal et al. 2009b)

scale τ will be discussed below). Using a positioning error in the range 150–500 m and $\tau = 3$ h for the IABP data, one finds an accuracy of $\delta U = \frac{\sqrt{2}\delta X}{\tau} \approx 2\text{–}6$ cm/s for the ice speed. In the mid-1990s, GPS started to replace ARGOS as the positioning system for some buoys, with an improved accuracy (down to 20–30 m Heil et al. 2008).

Satellite estimates of ice velocity are based on several kinds of sensors, such as synthetic aperture radar (SAR), radar scatterometers, or microwave radiometers. The basic tool to estimate displacements, and so velocities, over the ice cover is based on area correlation techniques between successive images of the same field (e.g. Ezraty et al. 2006; Fily and Rothrock 1987). The principal limitation of such analysis is that it measures spatial lags, and therefore very poorly retrieves the rotational component of ice motion. More sophisticated algorithms based on a combination of area-correlation techniques and feature-matching techniques have been developed and applied to SAR imagery (Kwok et al. 1990), furnishing the RGPS dataset (Kwok 1998). This dataset gives Lagrangian tracking of more than 40000 points over the Arctic basin, separated on average by 10 km.

The velocity accuracy obtained from classical correlation techniques applied to radiometers or SAR images ranges from several mm/s to several cm/s, for a temporal sampling of 2–3 days and a spatial resolution varying from 5 to 60 km, depending on the sensor/satellite (Rozman et al. 2011). The RADARSAT dataset shows better results, with a position accuracy of ~ 300 m (Lindsay and Stern 2003). This gives a velocity error of ~ 2 mm/s, but still for a temporal sampling of 3 days on average.

From these numbers, one realizes that satellites are unrivaled for the analysis of sea ice drift patterns with regular spatial and temporal sampling, but are strongly limited in terms of time resolution, i.e. they cannot be used to explore velocity fluctuations at "small" time scales. In this case, the analysis of buoy trajectories is required, at the cost of irregular spatial sampling.

2.2 How to Extract a Mean Field

We start here with a quotation from Thorndike (1986a): "The motion of a particle of sea ice can be partitioned into a predictable component, associated with the long-term average wind and ocean currents, and a random part associated with the short-term fluctuations in the wind and current". This suggests (i) the existence of an Arctic general circulation (AGC), and (ii) that the remaining fluctuating part of the motion simply results from oceanic and atmospheric turbulence.

Such a general circulation was already postulated by Nansen, and the Fram's journey confirmed his guess about a transpolar current. Almost one century later, Colony and Thorndike (1984) published the first estimate of the AGC from a compilation of trajectories, from the Fram's cruise to the first trajectories of automatic ice drifters (1979–1982). The main drawback of this dataset was its very

2.2 How to Extract a Mean Field

large spatial and temporal sampling variability. Therefore, Colony and Thorndike (1984) used an optimal linear interpolation procedure to spatially smooth the sparse data, and calculated a mean field for a time interval spanning 90 years. The resulting field shows two distinct features (Fig. 2.2): an anti-cyclonic circulation in the western Arctic, the so-called Beaufort Gyre, and the transpolar drift in the Eurasian basin. This mean field was then considered as the AGC, linked to first order to the large-scale atmospheric circulation (Colony and Thorndike 1984). The role of this circulation on the mass balance of Arctic sea ice, and so on its long-term evolution, is essential, especially through the sea ice export across Fram Strait (e.g. Kwok and Rothrock 1999; Kwok et al. 2004; Rampal et al. 2011). Consequently, many authors discussed a possible correlation between the inter-annual variability of atmospheric circulation and the evolution of sea ice concentration over the Arctic (e.g. Maslanik et al. 2007; Rigor et al. 2002). In so doing, they implicitly assumed that the AGC depicted in Fig. 2.2 might be affected by large-scale, long-term variability. In other words, are the averaging scales used by Colony and Thorndike (1984) to extract their mean field (90 years and ~ 200 km) appropriate? Indeed, if these averaging scales are too large, homogenization will be too strong and small-scale/short-term variability of the mean circulation will be lost. Conversely, if they are too small, the AGC will include a stochastic component, biasing the analysis of velocity fluctuations as well as their causes.

To define averaging scales on a physical basis, Rampal et al. (2009b) developed an approach inspired from the study of fluid turbulence and its associated diffusion properties. Such an analogy was already suggested by Thorndike (1986a), and is based on a decomposition of the velocity field into a predictable (in a deterministic sense) mean "flow" and its fluctuations—the so-called "Reynolds decomposition".

The starting point is the turbulent diffusion theory of (Taylor 1921), who showed that in the case of steady and homogeneous turbulence without mean flow, the diffusion of single particles through time, measured by the variance of the distance from the origin $\left\langle (X'(t) - X'(0))^2 \right\rangle = \left\langle r'^2(t) \right\rangle$, is linked to the variance of

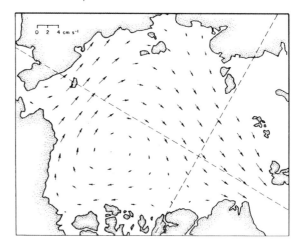

Fig. 2.2 An estimate of the mean field of Arctic sea ice motion obtained by Colony and Thorndike (1984) from an interpolation of various manned research stations or buoy trajectories over the period 1893–1983

the speed fluctuation $\langle U'^2(t)\rangle$ and the normalized turbulent autocorrelation function $C(\tau)$ through:

$$\langle r'^2(t)\rangle = 2\Gamma\langle U'^2(t)\rangle \int_0^t C(\tau)d\tau \qquad (2.4)$$

where Γ is the so-called integral time (Taylor 1921). $C(\tau)$ is defined as:

$$C(\tau) = \frac{1}{\langle U'^2\rangle t_{max}} \int_0^{t_{max}} \mathbf{U}'(t)\mathbf{U}'(t+\tau)dt \qquad (2.5)$$

where $U' = U - \langle U\rangle$ is the velocity fluctuation and t_{max} the duration of data coverage. $C(\tau)$ decreases theoretically with increasing time lag, e.g. following an exponential decay, as the memory of previous displacements is lost beyond the characteristic memory time $\Gamma = \int_0^\infty C(\tau)d\tau$. This fundamental property can be used to check whether the mean flow has been correctly removed. Indeed, if the mean flow is not properly removed, residual autocorrelation will remain for very large time scales, a signature of its predictable character in the deterministic sense. Conversely, if the scales of averaging of the mean flow are too small, $C(\tau)$ becomes negative for intermediate time lags, as a random part of the motion is artificially included in the mean flow, and therefore removed from the velocity fluctuation field.

This methodology was applied to Arctic sea ice velocity by (Rampal et al. 2009b) from the IABP ice drifters dataset (Fig. 2.3). These authors calculated a velocity fluctuation field using variable time (T) and space (L) averaging scales for the mean circulation. To estimate the integral time $\Gamma^{L,T}$ corresponding to each pair of averaging scales, they integrated the autocorrelation function, averaged over all the buoy trajectories, up to the time of first zero crossing, which gives an upper bound on the true scale (Poulain and Niiler 1990) (see Rampal et al. 2009b for more technical details). Figure 2.3, corresponding to winter (November to May), shows that, as expected, when no mean field is removed, some autocorrelation remains after 20 days. A similar result was obtained previously by Thorndike (1986b) from a limited dataset. For $L < 400$ km and $T < 160$ days, the autocorrelation function shows negative values until approximately 20 days, and consequently $\Gamma^{L,T}$ underestimates the genuine integral time. $\Gamma^{L,T}$ increases with increasing L and T and stabilizes around 1.5 days for $L = 400$ km and $T = 160$ days, for which the autocorrelation function follows the theoretical "turbulent" behavior (Fig. 2.3). This integral time scale is similar to that found for isopycnal oceanic turbulence (Zhang et al. 2001).

We can therefore conclude that 400 km and 160 days are the appropriate averaging scales to define an AGC for sea ice in winter. A similar analysis performed for summer trajectories gives averaging scales of 200 km and 80 days, for

2.2 How to Extract a Mean Field

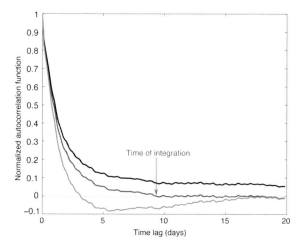

Fig. 2.3 Normalized autocorrelation functions of sea ice velocity fluctuations in winter. These velocities are obtained after removing a mean velocity field estimated from different spatial and temporal averaging scales. *Black curve*: no mean field removed. *Light gray*: mean field calculated for $L = 50$ km and $T = 40$ days. *Dark gray*: mean field calculated for $L = 400$ km and $T = 160$ days. The "time of integration" t_i indicates the time over which, in practice, the autocorrelation function $C(\tau)$ is integrated to estimate the integral time: $\Gamma = \int_0^{t_i} C(\tau)d\tau$ (from Rampal et al. 2009b)

an integral time of 1.3 days, slightly below the winter value (Rampal et al. 2009b). It is probably not fortuitous that the appropriate time scales correspond roughly in both cases to—actually are slightly below—the duration of one season, highlighting a seasonal variability of sea ice general circulation. These averaging scales also indicate an interannual variability of the AGC (Fig. 2.4), meaning that Colony and Thorndike's map of Fig. 2.2 represents an oversimplification. As an example, cyclonic circulation can be observed instead of the Beaufort Gyre (Fig. 2.4e). The only apparently persistent feature is the transpolar drift, although its magnitude and precise direction relative to Fram Strait may change from one year or season to another (compare Fig. 2.4a and b), with probable consequences in terms of sea ice export.

This variability of the AGC is likely associated with a change of forcing, as atmospheric and/or oceanic conditions exhibit a similar seasonal and interannual variability (e.g. Hurrell 1995; Thompson and Wallace 1998). It may have a strong impact on sea ice export and mass balance (see Conclusion and Perspectives, and Rampal et al. 2011). As an example, it has been argued that atmospheric circulation patterns favoring sea ice export to the northern Atlantic may have triggered the recent record lows of summer sea ice extent (Wang et al. 2009). However, the analysis presented above demonstrates that any correlation analysis between atmospheric circulation indexes, the sea ice AGC, and sea ice mass balance, should be performed using a sea ice circulation determined from the appropriate temporal and spatial scales, in order to remove any random fluctuation from the analysis.

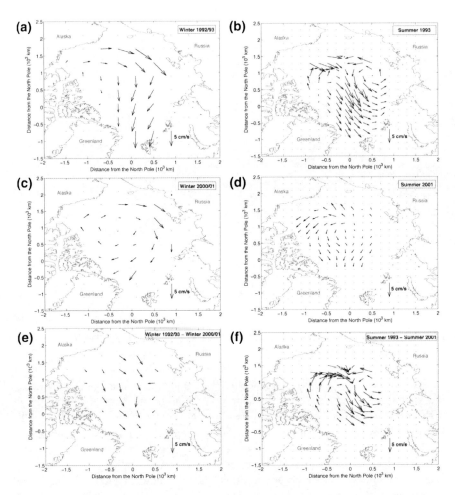

Fig. 2.4 Arctic general circulation (AGC) calculated from a non-arbitrary decomposition of the sea ice velocity field (see text and Rampal et al. 2009b for details) for different years and seasons. Compare with Colony and Thorndike's map of Fig. 2.2 (from Rampal et al. 2009b)

2.3 Diffusion Regimes

Once a general circulation has been determined at the appropriate scales, it can be removed from the velocity field in order to explore the fluctuating part of sea ice motion. The aim of this section is to analyze the statistical properties of these fluctuations in order to understand their physical origin, and particularly to check whether we can retrieve the fingerprint of oceanic or atmospheric turbulence, as suggested by Thorndike (1986a), and/or of another mechanism such as the intrinsic mechanical behavior of sea ice through the effect of internal stresses. To do so, it is interesting to build analogies with the Lagrangian statistics of fluid turbulence (e.g. Arneodo et al. 2008; Mordant et al. 2003).

2.3 Diffusion Regimes

The preliminary step, however, is to answer a question already formulated by Thorndike (1986b): Can a sea ice velocity be defined? The answer is not as trivial as it appears. Obviously, depending on the sampling interval of the trajectories, τ, it is always possible to define a velocity measured at this time scale, $\mathbf{U}'(\mathbf{X}, \tau) = \frac{\mathbf{X}(t+\tau) - \mathbf{X}(t)}{\tau} - \langle \mathbf{U} \rangle$ (we consider velocity fluctuations $\mathbf{U}'$ in this section). The pertinent question is to check whether a limit $\mathbf{U}'(\mathbf{X}, \tau \to 0) = \mathbf{U}'(\mathbf{X})$ exists and is approached for available time intervals, i.e. those longer than the sampling interval (3 h for the IABP data). The answer is related to the diffusion regimes of sea ice with regard to the turbulent diffusion theory of Taylor (1921) introduced in the previous section. From Eq. (2.4) and the definition of the integral time $\Gamma = \int_0^\infty C(\tau) d\tau$, it can be shown that particle diffusion $\langle r'^2(t) \rangle$ has two asymptotic regimes. For $t \ll \Gamma$, particle diffusion scales as $\langle r'^2(t) \rangle \sim t^2$: this is the so-called "ballistic regime" for which particles keep memory of previous velocity characteristics (magnitude as well as direction). For $t \gg \Gamma$, a regime corresponding to vanishing velocity correlations and "random walk" motion takes place, giving $\langle r'^2(t) \rangle \sim t$, as for molecular diffusion.

A condition for the convergence of $\mathbf{U}'(\mathbf{X}, \tau \to 0)$ is that the associated variance $\langle (\mathbf{U}'(\mathbf{X}, \tau \to 0))^2 \rangle$ converges as well (Thorndike 1986b). As $\langle \mathbf{U}'^2 \rangle \sim \langle r'^2(t) \rangle / t^2$, this condition consists merely of the existence of a ballistic regime for sea ice—in simpler words, a velocity is well defined for a ballistic trajectory. Figure 2.5, extracted from (Rampal et al. 2009b), shows sea ice diffusive displacements r'_x and r'_y (along the x and y directions of a Cartesian coordinate system centered on the North Pole with the y axis following the Greenwich meridian) caused by the velocity fluctuation field, in winter and summer. The symmetry about the zero line indicates that the mean circulation was correctly determined and removed (see Sect. 2.2). The shape of the envelope of the buoy trajectories qualitatively follows Taylor's diffusion predictions, with a rapid growth during the first few days followed by a slowing down. This is quantitatively confirmed in Fig. 2.6 where the two diffusion regimes are clearly identified, with a crossover at the integral time scale $\Gamma \approx 1.5$ days. This implies that a true velocity can be defined for sea ice, and is correctly estimated at the IABP time resolution of 3 h, as already suggested by Thorndike (1986b).

In their global analysis, Rampal et al. (2009b) did not distinguish between zonal and meridional components of motion. Such distinction was made by Lukovich et al. (2011) in a regional study of sea ice drift on the seasonal sea ice zone of the southern Beaufort Sea, north of Alaska, in winter. From this limited dataset (20 buoys), these authors reported two distinct diffusion regimes, a ballistic regime in t^2 for zonal motion, and a possible "intermediate" regime in $t^{5/4}$ for meridional motion along the so-called circumpolar flaw lead, both observed over a time range of 2–200 days. From an autocorrelation analysis of sea ice speed, they obtained an integral time scale Γ slightly larger for zonal motion (1.2 days) than for meridional motion (0.7 days). They interpreted the ballistic zonal motion observed at

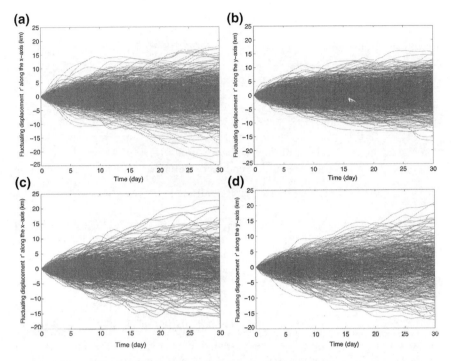

Fig. 2.5 Buoy displacements r'_x (**a** and **c**) and r'_y (**b** and **d**) caused by the sea ice velocity fluctuation field for winter (**a** and **b**) and summer (**c** and **d**). This illustrates the diffusion regimes of sea ice: a "rapid" initial ballistic regime ($\langle r'^2(t)\rangle^{1/2} \sim t$) followed by a molecular diffusion–like regime ($\langle r'^2(t)\rangle^{1/2} \sim t^{1/2}$) (from Rampal et al. 2009b)

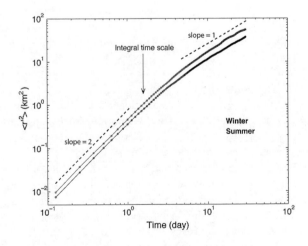

Fig. 2.6 Diffusion regimes of sea ice motion fluctuations obtained from the time scaling of the variance $\langle r'^2(t)\rangle$ of displacement fluctuations r'. The initial ballistic regime $\langle r'^2(t)\rangle \sim t^2$ is obtained for $t < \Gamma = 1.5$ days, whereas a random walk regime is obtained for $t > \Gamma$. (from Rampal et al. 2009b)

2.3 Diffusion Regimes

timescales much longer than the corresponding integral timescale as a result of advection-dominated flow, but one may question whether the mean field was correctly removed (see Sect. 2.2). Following Elhmaidi et al. (1993), they argued that the $\langle r'^2(t) \rangle \sim t^{5/4}$ anomalous meridional diffusion could be a signature of a deformation-dominated (shear and divergence) regime. Although the distinction between a $t^{5/4}$ and a classical linear regime is probably difficult to detect with a limited dataset, the possibility of such anomalous diffusion for sea ice would be worth pursuing in the future for a better understanding of the underlying processes.

2.4 Turbulent-Like Fluctuations

Once we know that the velocity fluctuations are correctly measured from ice drifter trajectories, and the diffusion regimes determined, further statistical analyses can be performed, and analogies with fluid turbulence pursued. The classical turbulent diffusion theory assumes a Gaussian distribution of velocities (Batchelor 1960; Frisch 1995; Taylor 1921). Eulerian wind velocities (i.e. the velocities "felt" by an ice parcel moving at a much lower speed than the wind speed) are Gaussian (Fig. 2.7). Lagrangian oceanic velocities are also Gaussian, once the mean flow has been correctly removed (Swenson and Niiler 1996; Zhang et al. 2001), although some non-Gaussian deviations have been claimed (Bracco et al. 2000). If one assumes that sea ice velocity fluctuation statistics are a direct inheritance of oceanic and atmospheric turbulence (see the beginning of this chapter and Thorndike 1986a), one would expect Gaussian statistics for sea ice as well. The reality is significantly different (Rampal et al. 2009b): sea ice velocity PDFs are exponential instead of Gaussian, with a few outliers above 60 cm/s (Fig. 2.8). This means that sea ice velocities fluctuate more (are more spread out),

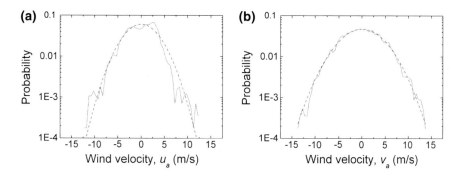

Fig. 2.7 Probability density functions (PDF) of the u_a (*left*) and v_a (*right*) components of the wind velocity vector $\mathbf{U_a}$ measured from November 1997 to September 1998 at an altitude of ~ 10 m on the meteorological tower of the SHEBA camp in the Beaufort Sea (see Persson et al. (2002) for more details on the measurements and Perovich et al. (1999) for the SHEBA program). The corresponding Gaussian distributions (*red dashed lines*) are given for comparison

in relative value, than wind or currents, and suggests that another term, namely sea ice mechanics and the internal stress, is playing a significant role.

To go further, the intermittency of the motion should be analyzed. This can be done through the analysis of the statistics of velocity fluctuation increments $\Delta_\tau \mathbf{U}' = \mathbf{U}'(t+\tau) - \mathbf{U}'(t)$ for different time scales τ (e.g. Mordant et al. 2003). In fluid turbulence, the PDFs of Lagrangian velocity increments continuously change as τ decreases within the inertial range of motion towards the Kolmogorov dissipative time scale τ_η. They are Gaussian at large time scales, above the integral time Γ, in agreement with Gaussian statistics of velocities and the absence of correlation between successive velocity values at these time scales. As τ decreases, the PDFs widen towards a stretched exponential form, $P(\mathcal{X}) \sim \exp\left(-\left(\frac{\mathcal{X}}{\mathcal{X}_0}\right)^\alpha\right)$ with $\alpha < 1$ (Mordant et al. 2003), in agreement with stretched exponential statistics for Lagrangian accelerations below τ_η (La Porta et al. 2001; Volk et al. 2008). Indeed, the acceleration estimated at time scale τ is given by $\left(\frac{\Delta_\tau \mathbf{U}'}{\tau}\right)$, and stabilizes below the dissipative time scale τ_η, i.e. a genuine acceleration is correctly approached for $\tau \ll \tau_\eta$. A similar analysis has been performed by Rampal et al. (2009b) for IABP velocity increments (Fig. 2.9). As for fluid turbulence, the PDF continuously changes as the time scale τ decreases, but the shapes of the distributions are significantly different. For $\tau > \Gamma$, the PDF is exponential and independent of τ, in agreement with exponential velocity statistics (see Fig. 2.8) and the absence of correlation between successive velocity values in this case. The variability of velocity increments increases towards small time scales ($\tau \ll \Gamma$), as the PDF develops power law tails that depart significantly from a stretched exponential form (Rampal et al. 2009b). In other words, the variability of sea ice velocity increments is even more extreme than that of fluid turbulence. This is another indication that the fluctuating part of sea ice motion is not (only) a direct consequence of wind or oceanic turbulence.

The widening of the velocity increment PDFs is an indication of a temporal "localization" of velocity fluctuations at short time scales, i.e. of intermittency.

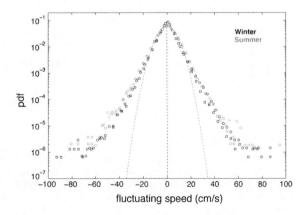

Fig. 2.8 PDFs of the u' (circles) and v' (squares) components of the sea ice velocity fluctuation vector $\mathbf{U}'$. The PDFs are symmetrical and show an exponential form that departs significantly from the corresponding Gaussian distribution (plotted for winter as the *gray dashed line*). (from Rampal et al. 2009b)

2.4 Turbulent-Like Fluctuations

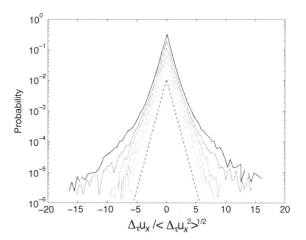

Fig. 2.9 PDFs of the sea ice velocity fluctuation increment $\Delta u'$ (x-component, normalized by its standard deviation) calculated over different time intervals τ. From top to bottom: $\tau = 0.125, 0.25, 0.5, 1, 2, 5$ days. The *dashed line* shows an exponential decay and is for reference only (from Rampal et al. 2009b)

A quantification of this intermittency can be done from the scaling of the structure functions:

$$\langle \Delta_\tau \mathbf{U}'^q \rangle \sim \tau^{\zeta(q)} \tag{2.6}$$

Figure 2.10 shows the structure functions $\langle |\Delta_\tau u'|^q \rangle$ of the y-component of Arctic sea ice fluctuating velocity, u' (similar results are obtained for the y-component, v'), for $0.5 \leq q \leq 3$. Scaling (Eq. 2.6) is observed from the time resolution of the IABP dataset (3 h) to $\tau \approx \tilde{\Gamma}$, which defines the upper bound of the so-called inertial range for turbulent flows. The slopes of Fig. 2.10 define the function $\zeta(q)$, which is non-linear (Rampal et al. 2009b). This nonlinearity is a signature of multifractality in the time domain, i.e. of intermittency, in agreement with the functional change in the PDFs (Fig. 2.9), and is remarkably well approximated by a quadratic fit $\zeta(q) = aq^2 + bq$ with $a = -0.09$ and $b = 0.65$.

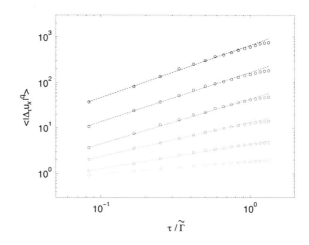

Fig. 2.10 Lagrangian structure functions of the x-component of the sea ice fluctuating velocity for $q = 0.5, 1, 1.5, 2, 2.5$ and 3 for winter. The *dashed lines* are the least squares power law fits $\langle \Delta_\tau u'^q \rangle \sim \tau^{\zeta(q)}$, yielding the exponents $\zeta(q)$ (from Rampal et al. 2009b)

Multifractality is a well-known property of turbulent flows at high Reynolds numbers (Frisch 1995; Mordant et al. 2003; Schmitt et al. 1992). The results of Rampal et al. (2009b) show that it also occurs for a solid, the Arctic sea ice cover, although the distribution of velocity increments is even more spread out in the case of sea ice.

2.5 Sea Ice Acceleration and the Dynamical Origin of Intermittency

We have seen above that sea ice motion shares several properties with fluid turbulence, such as diffusion regimes and intermittency, i.e. sea ice velocity fluctuations are "turbulent-like". On the other hand, we also noted important differences, particularly stronger variability and intermittency in the case of sea ice. As already stressed, this precludes a simple inheritance from atmospheric and/ or oceanic turbulence, contrary to what Thorndike (1986a) suggested. This raises the question of a possible dynamical origin of this intermittency, through the sea ice momentum equation (Sect. 1.2) that links sea ice acceleration to wind forcing, ocean drag, and the ice internal stress field.

We are faced here with a question similar to the one asked at the beginning of Sect. 2.3 for velocity: Can a sea ice acceleration be defined from available measurements? Once again, Thorndike (1986b) already asked this question 25 years ago. As already mentioned above, ice acceleration is obtained from $\left(\frac{\Delta_\tau U'}{\tau}\right)_{\tau \to 0}$, and therefore the associated variance $\left\langle \left(\frac{\Delta_\tau U'}{\tau}\right)^2 \right\rangle_{\tau \to 0}$ should stabilize above the time resolution of the measurements in order to correctly estimate the acceleration. The analysis of the structure functions in Fig. 2.10 shows that $\left\langle \left(\frac{\Delta_\tau U'}{\tau}\right)^2 \right\rangle \sim \frac{1}{\tau^2}\langle\Delta_\tau U'^2\rangle \sim \tau^{\zeta(2)-2} = \tau^{-1.06}$, i.e. the variance of acceleration increases towards short time scales down to the time resolution of the IABP data. In other words, the equivalent of the Kolmogorov dissipative time scale of fluid turbulence, τ_η, for sea ice motion is well below 3 h, with the consequence that ice acceleration is not correctly approached at this time resolution (note however that such acceleration is necessarily defined at sufficiently small time scales).

Nevertheless, from the analysis of velocity increment PDFs detailed in the preceding section, we expect a very strong intermittency of the true sea ice acceleration, with a PDF characterized by heavy, power law tails. Indeed, such heavy tails are already present at $\tau = 3$ h (Fig. 2.9) and should even widen below 3 h. This is in agreement with estimated accelerations obtained by Chmel et al. (2007) from the GPS trajectory of the North-Pole Station 32 sampled at a time resolution of 10 min. At such small time scale, these authors reported a power law distribution of accelerations, $P(A_i) \sim A_i^{-\alpha}$ with $\alpha \approx 4$.

2.5 Sea Ice Acceleration and the Dynamical Origin of Intermittency

Can such power law statistics be expected from simple arguments based on the momentum equation of sea ice (Eq. 1.1)? As discussed in Sect. 1.2, the main driving force of sea ice motion is the wind stress, which scales as $\tau_a \sim |\mathbf{U_a}|\mathbf{U_a}$. Wind speeds are normally distributed (Fig. 2.7 and Frisch 1995), and if a random variable $\mathcal{X}$ follows a standard normal distribution, the variable $\mathcal{Y} = |\mathcal{X}|\mathcal{X}$ will follow a χ^2 distribution with one degree of freedom, with a PDF given by:

$$P(\mathcal{Y}) = \frac{1}{\sqrt{2\pi|\mathcal{Y}|}} \exp\left(-\frac{|\mathcal{Y}|}{2}\right) \tag{2.7}$$

This distribution exhibits exponential tails: while more spread out than Gaussian statistics, the distribution of wind stresses cannot explain the extreme variability of sea ice accelerations. Once again, another source of variability and intermittency should be sought.

As shown above, the intermittency of sea ice kinematics implies that an estimate of acceleration, $\left(\frac{\Delta_\tau \mathbf{U}'}{\tau}\right)$, increases with decreasing time scale τ, and that the true value cannot be approached from buoy position measurements with a sampling frequency of a few hours. This raises the interesting question of the importance of the inertial forces associated with acceleration in the momentum balance of the ice (Thorndike 1986b). It is generally believed that these forces are very small compared to the other terms of the force balance (e.g. Coon 1980; Hibler 1984; Leppäranta 2005), thus suggesting the use of a "quasi-static" approximation of (Eq. 1.1) to estimate the amplitude of the internal stress term as a residual of the force balance (e.g. Rothrock et al. 1980). This assumption is valid when considering large time scales (>1 day), i.e. it is relevant for most applications and modeling issues. However, assuming that the scaling properties of sea ice velocity increments detailed in the previous section can be extrapolated below $\tau = 3$ h, at which time scale will the inertial forces become comparable to the other terms? To estimate this, the statistical and scaling properties of the ice velocity fluctuation field can be used. Using a mean speed $\langle U_i' \rangle = 7.8$ cm·s^{-1} (Rampal et al. 2009b), $f \approx 1.4 \times 10^{-4}$ s^{-1}, a thickness $h = 1$ m and $\rho_i = 910$ kg·m^{-3} (Timco and Frederking 1996), one finds a Coriolis "force" (per unit area) $\rho h f \langle U_i' \rangle \approx 10^{-2}$ Pa. A typical average surface wind speed $\langle U_a \rangle$ is 5 m·s^{-1}, although values larger than 10 m·s^{-1} are observed (see Fig. 2.7). This leads to a wind stress $\tau_a = \rho_a C_a \langle U_a \rangle^2 \times \approx \times 5 \cdot 10^{-2}$ Pa, taking $\rho_a = 1.3$ kg·m^{-3} and $C_a = 1.5 \times 10^{-3}$ (Leppäranta 2005). For the IABP time resolution $\tau = 3$ h, a typical incremental velocity $\left\langle (\Delta U')^2 \right\rangle^{1/2}_{\tau=3h}$ is about 2 cm·s^{-1}, giving an inertial "force" term of $\rho h \frac{\langle (\Delta U')^2 \rangle^{1/2}_\tau}{\tau} \approx 1.7 \times 10^{-3}$ Pa. This is still well below the wind stress or the Coriolis term, in agreement with the quasi-static assumption noted above. However, if ones extrapolates the scaling $\frac{1}{\tau}\langle \Delta U'^2 \rangle^{1/2}_\tau \sim \tau^{-0.53}$ below 3 h, one finds that the inertial force becomes comparable to the Coriolis term for a time scale $\tau \approx 400$ s, and to the wind stress for $\tau \approx 20$ s. The question that arises is whether the multifractality of sea ice motion indeed holds down to such small time scales, or if a lower bound to the inertial range of sea ice motion, corresponding to some equivalent of the Kolmogorov time scale of dissipation, is attained before. Owing to the propagation of errors when estimating accelerations from buoy positions, it is questionable whether one could explore this issue from such data. Other kinds of field instrumentation, such as seismometers (e.g. Marsan et al. 2011) or accelerometers, might be an interesting alternative in the future.

2.6 Spectral Analysis

So far, we have focused on sea ice kinematics in the time domain. A complementary approach is a spectral analysis in the frequency domain. From a linear theoretical analysis of free drift, i.e. using a linearized drag coefficient, Leppäranta et al. (2012) recently showed that the Lagrangian power spectrum of sea ice velocity $p_i(\omega)$ is expressed in this ideal case as the combination of the power spectrum of the forcing $p_F(\omega)$ and of a modulation factor $a(\omega, f)$ that can be understood as the sea ice spectral response to white noise, $p_i(\omega) = a(\omega, f) p_F(\omega)$, where ω is the frequency. This model exhibits a singularity at the Coriolis frequency $\omega = -f$, a signature of inertial oscillations. The modulation factor tends towards a constant as $\omega \to 0$ and is essentially flat for $\omega < 0.38$ cycles·day^{-1}, meaning that the sea ice velocity spectrum simply follows the forcing spectrum in this range. Towards high frequencies, and typically for $\omega > 1$ cycle·h^{-1}, $a(\omega, f) \sim \omega^{-2}$ (Leppäranta et al. 2012).

Rampal et al. (2009b) calculated the Lagrangian power spectrum of the IABP velocity fluctuations $\mathbf{U}_i'$ from a Fourier transform of the autocorrelation function (see Sect. 2.2). Obviously, the time resolution of the IABP dataset does not allow one to explore the high frequency range. Rampal et al. (2009b) reported a Kolmogorov-like scaling $p_i(\omega) \sim \omega^{-2}$ over the frequency range 0.1–2 cycle·day^{-1}. For lower frequencies corresponding to time scales significantly larger than the integral time scale Γ, the scaling falls off, in agreement with the white noise character of the sea ice velocity in this case (see Sect. 2.3), whereas a clear inertial peak is observed at $\omega = -f \approx 2$ cycles·day^{-1}. Sea ice inertial oscillations have been reported and studied by many authors (e.g. Colony and Thorndike 1980; Heil and Hibler 2002; Hunkins 1967; McPhee 1978; Thorndike 1986b). The associated inertial frequency peak is less pronounced in winter and/or within the central pack, compared to summer or peripheral zones of the Arctic basin, a signature of damping of these oscillations by ice internal stresses (Gimbert et al. 2012b).

The $p_i(\omega) \sim \omega^{-2}$ scaling reported by Rampal et al. (2009b) cannot be explained entirely by the high frequency asymptotic behavior of the free drift model of Leppäranta et al. (2012), as it extends over much lower frequencies. It might be instead a signature of the forcing spectrum. Indeed, such ω^{-2} Lagrangian scaling has been observed for oceanic (Lien et al. 1998) and atmospheric turbulence

(Gifford 1956; Hanna 1981). In this respect, the free drift model that predicts $p_i(\omega) = a \times p_F(\omega)$ for $\omega < 0.38$ cycle·day^{-1}, where a is a constant, gives a correct description of sea ice drift in the spectral domain, and Thorndike's proposition of a fluctuating part of ice motion simply resulting from oceanic and atmospheric turbulence would appear reasonable. Such spectral analysis furnishes however only a limited view of the complexity of sea ice motion. Indeed, from the $p_i(\omega) \sim \omega^{-2}$ scaling, we expect that the second order structure function scales as $\left\langle \Delta_\tau U'^2 \right\rangle \sim \tau$ (Biferale et al. 2006; Mordant et al. 2003), in agreement with the value $\zeta(2) = 0.95$ (see Rampal et al. 2009b) and Sect. 2.4), but the power spectrum scaling tells nothing about the multifractality of velocity fluctuations that fully characterizes the statistics.

Finally, we note that the Lagrangian spectrum of IABP sea ice velocity increments calculated at $\tau = 3$ h is flat, whereas Chmel and Smirnov (2007) reported a correlation time scale for their NP-32 acceleration time series of a few tens of minutes at most. This means that the fluctuating part of sea ice motion can be considered as a Markov process down to a (still unresolved) dissipative time scale that is much shorter than 1 h.

2.7 Concluding Remarks

In this chapter, we have explored the statistical and scaling properties of sea ice drift from buoy trajectories (Lagrangian tracers). The first important conclusion is that the velocity field of sea ice motion can be decomposed non-arbitrarily into a mean field and turbulent-like fluctuations. The properties of the remaining velocity fluctuation field have been explored in the time and frequency domains, in which they share similarities with fluid turbulence, such as a Kolmogorov-like scaling of the power spectral density, the presence of a ballistic regime within an inertial range of motion (whose lower bound is however not yet resolved), and the presence of intermittency. This may suggest a rather direct inheritance from the properties of atmospheric and oceanic turbulence, i.e. a linear response of sea ice to the forcing fields. However, a closer look indicates significant differences between geophysical fluid turbulence and sea ice motion: velocity PDFs are clearly not Gaussian, with much more spread; intermittency is more pronounced; and the distributions of sea ice accelerations cannot be explained by wind stress statistics. These differences are likely the signature of a non-linear response of sea ice to the forcing fields.

The kinematic response of a solid to forces and stresses occurs through a rheology F linking internal stress σ to strain ε and/or strain-rate $\dot{\varepsilon}$, $\sigma = F(\varepsilon, \dot{\varepsilon})$. If the properties detailed in this chapter suggest a non-linear character for F, an analysis of sea ice strains and strain-rates, i.e. velocity fluctuations and gradients in the spatial domain, is essential to go further.

References

Arneodo, A., et al. (2008). Universal intermittent properties of particle trajectories in highly turbulent flows. *Physical Review Letters, 100*(25), 254504.

Batchelor, G. K. (1960). *The theory of homogeneous turbulence*. Cambridge: Cambridge University Press.

Biferale, L., Boffetta, G., Celani, A., Lanotte, A., & Toschi, F. (2006). Lagrangian statistics in fully developed turbulence. *Journal of Turbulence, 7*(6), 1–12.

Bracco, A., LaCasce, J. H., & Provenzale, A. (2000). Velocity probability density functions for oceanic floats, *J. Physical Oceanography, 30*, 461–474.

Chmel, A., & Smirnov, V. (2007). The artic-sea-ice cover: problem of forecasting. *Physica a-Statistical Mechanics and Its Applications, 375*(1), 288–296.

Chmel, A., Smirnov, V. N., & Panov, L. V. (2007). Scaling aspects of the sea-ice-drift dynamics and pack fracture. *Ocean Science, 3*(2), 291–298.

Colony, R., & Thorndike, A. S. (1984). An estimate of the mean field of the arctic sea ice motion, *Journal of Geophysical Research, 89*(C6), 10623–10629.

Colony, R., & Thorndike, A. S. (1980). The horizontal coherency of the motion of summer Arctic sea ice. *The Journal of Physical Oceanography, 10*, 1281–1289.

Coon, M. (1980). A review of AIDJEX modeling, in *Sea ice processes and models*, Pritchard, R. S. (Ed.), (pp. 12–25). University of Washington Press.

Elhmaidi, D., Provenzale, A., & Babiano, A. (1993). Elementary topology of 2-dimensional turbulence from a lagrangian viewpoint and single-particle dispersion. *Journal of Fluid Mechanics, 257*, 533–558.

Ezraty, R., Ardhuin, F., & Piollé, J. F. (2006). *Sea ice drift in the central arctic estimated from Seawinds/Quickscat backscatter maps - User's manual*. Brest: Rep IFREMER/CERSAT.

Fily, M., & Rothrock, D. A. (1987). Sea ice tracking by nested correlations. *IEEE Geoscience and Remote Sensing Letters, 25*(5), 570–580.

Frisch, U. (1995). *Turbulence The legacy of A.N Kolmogorov*. Cambridge: Cambridge University Press.

Gifford, F. (1956). A simultaneous lagrangian-eulerian turbulence experiment. *Monthly Weather Review, 83*(12), 293–301.

Gimbert, F., Jourdain, N. C., Marsan, D., Weiss, J. & B. Barnier (2012b), Recent mechanical weakening of the Arctic sea ice cover as revealed from larger inertial oscillations, *Journal of Geophysical Research, 117*, C00J12.

Hanna, S. R. (1981). Lagrangian and eulerian time-scale relations in the daytime boundary layer. *Journal of Applied Meteorology, 20*, 242–249.

Heil, P., & Hibler, W. D. I. (2002). Modeling the high frequency component of arctic sea ice drift and deformation. *The Journal of Physical Oceanography, 32*, 3039–3057.

Heil, P., Hutchings, J. K., Worby, A. P., Johansson, M., Launiainen, J., Haas, C., et al. (2008). Tidal forcing on sea-ice drift and deformation in the western Weddell Sea in early austral summer, 2004. *Deep-Sea Research Part II Topical Studies in Oceanography, 55*(8–9), 943–962.

Hibler, W. D. I. (1984). Ice dynamics, *Rep. 84–83*, CRREL.

Hunkins, K. (1967). Inertial oscillations of Fletcher's ice island (T-3). *Journal of Geophysical Research, 72*(4), 1165–1174.

Hutchings, J. K., Heil P., Steer A., & Hibler W. D. (2012). Subsynoptic scale spatial variability of sea ice deformation in the western Weddell Sea during early summer, *Journal of Geophysical Research-Oceans, 117*, C01002.

Hurrell, J. W. (1995). Decadal trends in the north-atlantic oscillation - regional temperatures and precipitation. *Science, 269*(5224), 676–679.

Kwok, R., Curlander, J. C., McConnell, R., & Pang, S. S. (1990). An ice-motion tracking system at the Alaska SAR facility. *IEEE Journal of Oceanic Engineering, 15*(1), 44–54.

Kwok, R. (1998). The RADARSAT geophysical processor system, in *Analysis of SAR data of the polar oceans*, Tsatsoulis, C. & Kwok, R. (Ed.), (pp. 235–257). Springer-Verlag.

Kwok, R., & Rothrock D. A. (1999). Variability of fram strait ice flux and north Atlantic oscillation, *Journal of Geophysical Research, 104*(C3), 5177–5189.

Kwok, R., Cunningham, G. F., & Pang, S. S. (2004). Fram Strait sea ice outflow. *Journal of Geophysical Research, 109*, C01009.

La Porta, A., Voth, G. A., Crawford, A. M., Alexander, J., & Bodenschatz, E. (2001). Fluid particle accelerations in fully developed turbulence. *Nature, 409*(6823), 1017–1019.

Lien, R. C., D'Asaro, E. A., & Dairiki, G. T. (1998). Lagrangian frequency spectra of vertical velocity and vorticity in high-Reynolds-number oceanic turbulence. *Journal of Fluid Mechanics, 362*, 177–198.

Leppäranta, M. (2005). *The drift of sea ice*. Berlin: Springer.

Leppäranta, M., Oikkonen, A., Shirazawa, K., & Fukamachi, Y. (2012). A treatise on frequency spectrum of drift ice velocity. *Cold Regions Science and Technology, 76–77*, 83–91.

Lindsay, R. W., & Stern, H. L. (2003). The RadarSat geophysical processor system: quality of sea ice trajectory and deformation estimates. *Journal of Atmospheric and Ocean Technology, 20*, 1333–1347.

Lukovich, J. V., Babb, D. G. & Barber, D. G. (2011). On the scaling laws derived from ice beacon trajectories in the southern Beaufort Sea during the international polar year - circumpolar flaw lead study, 2007–2008, *Journal of Geophysical Research-Oceans, 116*.

Marsan, D., Weiss, J., Metaxian, J. P., Grangeon, J., Roux, P. F., & Haapala, J. (2011). Low-frequency bursts of horizontally polarized waves in the Arctic sea-ice cover. *Journal of Glaciology, 57*(202), 231–237.

Maslanik, J., Dobrot, S., Fowler, C., Emery, W., & Barry, R. (2007). On the Arctic climate paradox and the continuing role of atmospheric circulation in affecting sea ice conditions. *Geophysical Research Letters, 34*, L03711.

McPhee, M. G. (1978). A simulation of inertial oscillation in drifting pack ice. *Dynamics of Atmospheres and Oceans, 2*, 107–122.

Mordant, N., Delour, J., Léveque, E., Michel, O., Arnéodo, A., & Pinton, J. F. (2003). Lagrangian velocity fluctuations in fully developed turbulence: scaling, intermittency, and dynamics. *Journal of Statistical Physics, 113*(5/6), 701–717.

Nansen, F. (1902). Oceanography of the north polar basin: The Norwegian north polar expedition 1893–96. *Scientific Results, 3*(9).

Perovich, D., et al. (1999). Year on ice gives climate insights. *EOS, 80*(41), 485–486.

Persson, P. O. G., Fairrall, C. W., Andreas, E. L., Guest, P. S. & Perovich D. K. (2002). Measurements near the atmospheric surface flux group tower at SHEBA: Near-surface conditions and surface energy budget, *Journal of Geophysical Research, 107*(C10), C000705.

Poulain, P. M., & Niiler P. P. (1990). The response of drifting buoys to currents and wind - comment, *Journal of Geophysical Research-Oceans, 95*(C1), 797–799.

Rampal, P., Weiss, J., Marsan, D., & Bourgoin, M. (2009b). Arctic sea ice velocity field: general circulation and turbulent-like fluctuations. *Journal of Geophysical Research, 114*, C10014.

Rampal, P., Weiss, J., Dubois, C., & Campin, J. M. (2011), IPCC climate models do not capture Arctic sea ice drift acceleration: Consequences in terms of projected sea ice thinning and decline, *Journal of Geophysical Research, 116*, C00D07.

Reed, R. J., & Campbell, W. J. (1962). The equilibrium drift of ice station Alpha. *Journal of Geophysical Research, 67*(1), 281–297.

Rozman, P., Holemann, J. A., Krumpen, T., Gerdes, R., Koberle, C., Lavergne, T., et al. (2011). Validating satellite derived and modelled sea-ice drift in the Laptev Sea with in situ measurements from the winter of 2007/08. *Polar Research, 30*, 7218.

Rigor, I. G., Wallace, J. M., & Colony, R. L. (2002). Response of sea ice to the Arctic oscillation. *Journal of Climate, 15*(18), 2648–2663.

Rothrock, D. A., Colony, R., & Thorndike, A. S. (1980). Testing pack ice constitutive law with stress divergence measurements. In R. S. Pritchard (Ed.), *Sea Ice Processes and Models* (pp. 102–112). Seattle: University of Washington Press.

Schmitt, F., Lavallée, D., Schertzer, D., & Lovejoy, S. (1992). Empirical determination of universal multifractal exponents in turbulent velocity fields. *Physical Review Letters, 68*, 305–308.

Swenson, M. S. & Niiler, P. P. (1996). Statistical analysis of the surface circulation of the California current, *Journal of Geophysical Research-Oceans, 101*(C10), 22631–22645.

Taylor, G. I. (1921). Diffusion by continuous movements. *The Proceedings of the London Mathematical Society, 20*, 196–212.

Thorndike, A. S. & Colony R. (1982). Sea ice motion in response to geostrophic winds, Journal of Geophysical Research, *87*(C8), 5845–5852.

Thomas, D. (1999). The quality of sea ice velocity estimates, *Journal of Geophysical Research, 104*(C6), 13627–13652.

Thorndike, A. S. (1986a). Kinematics of sea ice. In N. Untersteiner (Ed.), *The geophysics of sea ice* (pp. 489–549). New York: Plenum Press.

Thorndike, A. S. (1986b). Kinematics of sea ice, In: Untersteiner, N. (ed.) The geophysics of sea ice, Plenum Press, New York, 489–549.

Thompson, D. W. J., & Wallace, J. M. (1998). The Arctic Oscillation signature in the wintertime geopotential height and temperature fields. *Geophysical Research Letters, 25*(9), 1297–1300.

Timco, G. W., & Frederking, R. M. W. (1996). A review of sea ice density. *Cold Regions Science and Technology, 24*(1), 1–6.

Volk, R., Mordant, N., Verhille, G., & Pinton, J. F. (2008). Measurement of particle and bubble accelerations in turbulence. *Europhysics Letters, 81*, 34002.

Wang, J., Zhang, J., Watanabe, E., Ikeda, M., Mizobata, K., Walsh, J. E., et al. (2009). Is the dipole anomaly a major driver to record lows in Arctic summer sea ice extent? *Geophysical Research Letters, 36*, L05706.

Zhang, H. M., Prater, M. D. & Rossby T. (2001), Isopycnal lagrangian statistics from the North Atlantic current RAFOS float observations, *Journal of Geophysical Research, 106*(C7), 13817–13836.

Chapter 3
Sea Ice Deformation

Abstract Instead of individual trajectories, this chapter consider velocity gradients within sea ice, i.e. deformation rates. This can be done either from the analysis of the relative dispersion of tracers (buoys), or from strain-rate fields obtained from satellite imagery. Although statistical tools previously developed in the context of turbulence have been used to characterize sea ice dispersion, the differences between sea ice and turbulent fluids are important. Sea ice deformation exhibits a strong spatial localization accompanied by a strong intermittency, both aspects being respectively characterized by specific space and time scaling laws. Moreover, they are coupled together through a space/time scaling symmetry that has no equivalence in fluid turbulence, but has been documented for the brittle deformation of the Earth's crust. A consequence is that sea ice dispersion regimes vary from a slightly super-diffusive regime at small spatial scales, to a very slow sub-diffusive one at large scales, i.e. much slower than oceanic dispersion. This also argues for a strongly non-linear, brittle rheology of the sea ice cover.

In the preceding chapter, individual sea ice Lagrangian trajectories were analyzed with a particular focus on velocity fluctuations in the time domain. Velocity gradients, i.e. deformation rates, were not analyzed. Owing to the huge aspect ratio of lateral extension over thickness for the sea ice cover, we are essentially interested in the 2D strain and/or strain-rate fields in the horizontal plane. To estimate a 2D strain-rate tensor over a given spatial domain, at least three trajectories have to be recorded simultaneously, the square root of the enclosed area giving the spatial scale of measurement. As shown below, this scale of measurement is a critical parameter in the estimation of strain-rates, owing to the extreme localization of sea ice deformation. At a given time t, and knowing the sea ice velocity $\mathbf{U} = (u,v)$ at different positions $\mathbf{X} = (x,y)$, the strain-rate tensor can be calculated from the partial derivatives $\left(\frac{\partial u}{\partial x}, \frac{\partial u}{\partial y}, \frac{\partial v}{\partial x}, \frac{\partial v}{\partial y}\right)$ using standard finite difference formulas. In the analysis of mechanical processes and scaling properties, one generally focuses on the following invariants:

J. Weiss, *Drift, Deformation, and Fracture of Sea Ice*,
SpringerBriefs in Earth Sciences, DOI: 10.1007/978-94-007-6202-2_3,
© The Author(s) 2013

$$\begin{cases} \text{divergence}: \dot{\varepsilon}_{div} = \frac{\partial u}{\partial x} + \frac{\partial v}{\partial y} \\ \text{shear}: \dot{\varepsilon}_{shear} = \left[\left(\frac{\partial u}{\partial x} - \frac{\partial v}{\partial y}\right)^2 + \left(\frac{\partial u}{\partial y} + \frac{\partial v}{\partial x}\right)^2\right]^{1/2} \\ \text{total deformation}: \dot{\varepsilon}_{tot} = \sqrt{\dot{\varepsilon}_{div}^2 + \dot{\varepsilon}_{shear}^2} \end{cases} \quad (3.1\text{a}-\text{c})$$

The vorticity $\left(\frac{\partial u}{\partial y} - \frac{\partial v}{\partial x}\right)$, of primary importance in fluid mechanics, has been less studied in the case of sea ice. This might be due to its more limited interpretation in solid mechanics, although Thorndike and Colony (1982) reported a positive correlation between sea ice and geostrophic wind vorticities estimated at a large spatial scale (~ 500 km). In the sea ice literature, particular attention has been given to the divergence $\dot{\varepsilon}_{div}$, owing to its potential importance for ocean–atmosphere interactions along opening leads ($\dot{\varepsilon}_{div} > 0$) and new ice production (e.g. Stern et al. 1995), or for the formation of sea ice ridges in case of convergence, with a strong impact on the ice thickness distribution (Thorndike et al. 1975). For this reason, another invariant, $\phi = \arctan\left(\frac{\dot{\varepsilon}_{shear}}{\dot{\varepsilon}_{div}}\right)$, which indicates whether the deformation is predominantly divergent ($\phi \sim 0$), in shear ($\phi \sim \pi/2$), or convergent ($\phi \sim \pi$), has been sometimes used (Stern et al. 1995; Thorndike 1986; Thorndike et al. 1975). We can already note that (i) shear is the dominant deformation mode in the Arctic, i.e. the distribution of ϕ is strongly centered around $\pi/2$ (Stern et al. 1995), and (ii) shear and divergence are strongly correlated in space (Weiss and Schulson 2009), both observations supporting a sea ice deformation field mainly controlled by shear faulting associated with dilatancy (see Weiss and Schulson 2009 and Chap. 4).

3.1 Data

The kinematic data available to estimate sea ice deformation are basically the same as the ones used to analyze drift and velocities, i.e. ice drifters and satellite-derived velocities (see Sect. 2.1). The main limitation of ice drifters as a tool to calculate strain-rates is the small number of drifters running simultaneously, and with separations covering a large range of spatial scales, except when dense arrays of buoys are especially launched for this purpose (Hutchings et al. 2011, 2012). This limitation is reinforced by the fact that at least three trajectories are needed to calculate tensorial quantities, whereas four trajectories give much better estimates. A way to circumvent this is to consider the dispersion through time of pairs of simultaneous trajectories, much more numerous than triplets or quadruplets (Martin and Thorndike 1985; Rampal et al. 2008). This methodology, detailed in Sect. 3.3, does not, however, allow one to discriminate different deformation modes (shear vs. divergence).

When dense arrays of buoys are specifically launched, they offer unrivaled strain-rate estimates in term of time resolution and errors. The error on strain-rate

3.1 Data

$\delta\varepsilon$ is proportional to positioning error δX, and inversely proportional to the temporal sampling τ and to the average ice velocity U, $\delta\varepsilon \approx 2\sqrt{2}\frac{\varepsilon\delta X}{U\tau}$ (Hutchings et al. 2012). Another source of error is inversely related to the area A enclosed by the buoy array, scaling as $\frac{\varepsilon\delta X}{\sqrt{A}}$, but, for a sampling rate of 1 h and a typical ice velocity of 5 cm s^{-1}, this contribution becomes rapidly negligible compared to the former one for $\sqrt{A} > 1$ km. All of this means that the error is larger for small arrays moving slowly (Hutchings et al. 2011). Taking a GPS positioning error δX of 10–30 m and $U = 5$ cm s^{-1}, one finds a relative error $\delta\varepsilon/\varepsilon$ between 15 and 45 % for a time scale τ of 1 h. This reduces to 0.65–2.0 % for $\tau = 1$ day.

Another limitation of drifter arrays is that they cannot be dense at the regional or larger scales, for obvious logistical and financial reasons. Satellite imagery is therefore necessary to get a global view of sea ice deformation. We have seen in Sect. 2.1 that area correlation techniques between successive images poorly retrieve the rotational component of ice motion. This leads to strong biases when strain-rates are estimated by spatial differentiation of these velocity fields. Consequently, more sophisticated algorithms incorporating feature-matching techniques have been developed and applied to SAR imagery (Kwok et al. 1990). The corresponding RGPS dataset tracks more than 40000 points over a large part of the Arctic basin (Kwok 1998). This analysis can generally be performed from October to June; feature-tracking is more difficult in summer. These points initially (in October) define square 10×10 km cells over which the strain-rate tensor and corresponding invariants can be estimated from partial derivatives of the velocity field estimated between two successive satellite images of the same region. The repeat pass of the satellite therefore sets the time resolution of the computation, which is 3 days on average. This is the main limitation of this remarkable dataset that furnishes an unrivaled view of the deformation processes of a geophysical medium, the Arctic sea ice cover, from a local scale of 10 km to the scale of the Arctic basin (>1000 km). At the spatial (10 km) and temporal (3 days) resolution of the analysis, tracking errors give rise to error standard deviations of 5×10^{-3} day^{-1} in the divergence, shear, and vorticity (Lindsay and Stern 2003). Obviously, this error is reduced if one computes deformation over larger time and/or space windows, at the cost of reduced resolution. Using European Remote Sensing (ERS-1) SAR imagery with a spatial resolution of 100 m, Thomas et al. (2008) were able to extract high-resolution strain-rate fields at a 400 m resolution and an error standard deviation of 1.3×10^{-2} day^{-1}, for a temporal resolution of ~ 3 days in average. Compared to the RGPS dataset, this analysis qualitatively shows the increasing spatial localization of sea ice deformation towards small scales (see below), but corresponding datasets are not so far available to perform statistical analyses.

In this chapter, we review first in Sect. 3.2 several statistical and scaling analyses of the RGPS dataset that reveal the extreme localization of sea ice strain-rate patterns, when computed at a time scale of 3 days. As the time resolution of satellite imagery is too limited to allow a detailed analysis of the evolution of deformation in the time domain, this is performed in Sect. 3.3 from a study of the dispersion of pairs of drifters, in analogy with the dispersion of Lagrangian tracers in fluid turbulence. Spectral characteristics of sea ice deformation are briefly discussed in Sect. 3.6.

3.2 Spatial Scaling and Localization of Deformation

Figure 3.1 shows an example of divergence and shear rate fields in the Arctic Ocean for a 3-day period in January 1997 obtained from the RGPS dataset. The most striking features on these maps are the linear-like structures where shear as well divergence are localized, and that can span a large part of the Arctic basin. These structures, which have been sometimes called linear kinematic features (Kwok 2001; Moritz and Stern 2001), are actually active faults concentrating simultaneously shear and divergent deformation (Weiss and Schulson 2009). Indeed, considering a near 100 % ice concentration in winter, sea ice is confined within the Arctic basin and shear along rough faults is necessarily accompanied by dilatancy (e.g. Jaeger and Cook 1979), i.e. fault/lead opening. A preliminary illustration of the degree of strain localization is given by a simple number: at these resolution scales (3 days, 10 km), 50 % of all shear is accommodated by only 6 % of the sea ice area (Girard et al. 2009). Interestingly, this degree of strain localization is scale dependant, i.e. it decreases with increasing spatial scale. This is a first sign of the scaling properties of sea ice strain-rate patterns.

The next step is obtained from the analysis of the distributions of strain-rates. At the RGPS resolution scales, the PDFs of both shear and divergence rates exhibit power law tails, $P(\dot{\varepsilon}) \sim \dot{\varepsilon}^{-\eta}$, with $\eta \approx 2.4 \pm 0.1$ and without visible truncation towards large strain-rates (Girard et al. 2009; Marsan et al. 2004) (Fig. 3.2). This means that sea ice strain-rate distributions follow Levy's laws of stability parameter $\eta - 1$, characterized by "wild randomness" and dominated by extreme values. When the stability parameter is smaller than 2, that is $\eta < 3$, as obtained for strain rate distributions at 10 km, the standard deviation is, analytically, infinite (e.g. Sornette 2000). In practice, it is obviously always possible to compute an empirical standard deviation for a given set of strain rates, but the shape of the RGPS distributions implies that the few largest values have a substantial influence on the classical tools used to characterize fluctuations. This illustrates the extreme localization of sea ice deformation, in a sense now more clearly defined.

The distributions of strain-rates at some (arbitrary) scale are however not sufficient to characterize sea ice deformation. If the 3-day strain-rates of the 10 km cells were independent scalar variables identically distributed following Levy's law, the PDF of strain-rates at any larger scale would also be a Levy's law of the same parameter (e.g. Sornette 2000). Sea ice strain-rates show a slightly different picture, as the power law exponent η increases with increasing spatial scale (Girard et al. 2009; Marsan et al. 2004). This results from the fact that (i) strain-rates are tensorial variables and (ii) strong spatial correlations exist, i.e. cell strain-rates are not independent stochastic variables. A full analysis of the evolution of the

3.2 Spatial Scaling and Localization of Deformation 35

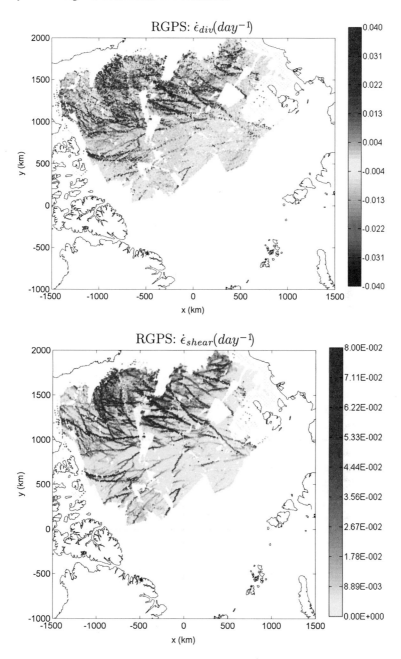

Fig. 3.1 Divergence (**a**) and shear (**b**) rate fields (in day^{-1}) from RGPS observations for the period 16–18 January 1997 (adapted from Girard et al. 2009)

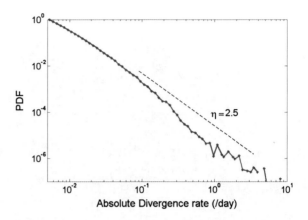

Fig. 3.2 PDF of absolute value of divergence rates for the period January–March 1997, from RGPS observations (adapted from Girard et al. 2009)

distributions with scale is therefore necessary. However, to determine the scaling symmetries directly from an analysis of the PDFs is, as noted in Sect. 1.3, difficult. A multifractal analysis of the moments $\langle \dot{\varepsilon}^q \rangle$ is more appropriate. Such analysis was first performed for the Arctic basin in winter (Marsan et al. 2004), and then extended over the annual cycle (Stern and Lindsay 2009) as well as, recently, to strain-rates estimated from an array of buoys in the Beaufort Sea (Hutchings et al. 2011). All these studies agree on the scale invariant character of the mean sea ice total strain-rate, i.e. the first order moment of the PDF (Fig. 3.3):

$$\langle \dot{\varepsilon}_{tot} \rangle \sim L^{-\beta(1)} \qquad (3.2)$$

where the exponent $\beta(1) = 0.18 \pm 0.10$ from November to April, and slightly increases during summer to reach $\beta(1) = 0.40 \pm 0.10$ in August (Stern and Lindsay 2009). Note that the standard deviations listed here (± 0.10) are from one 3-day

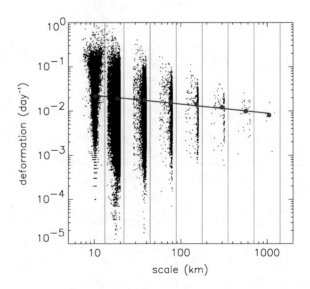

Fig. 3.3 Total strain-rate $\dot{\varepsilon}_{tot}$ as a function of scale L for one 3-days spatial pattern of Arctic sea ice deformation in November 1997. *Vertical lines* define bins, and *red dots* are average strain-rates $\langle \dot{\varepsilon}_{tot} \rangle_L$ within each bin. The blue solid line is least squares fit to these mean values, giving $\langle \dot{\varepsilon}_{tot} \rangle \sim L^{-0.20}$ (from Marsan et al. 2004)

3.2 Spatial Scaling and Localization of Deformation

RGPS snapshot to another, whereas the uncertainty on the estimation of β for a given deformation field is one order of magnitude smaller (Marsan et al. 2004). A confirmation of this scaling law, with a close exponent ($\beta(1) = 0.2 \pm 0.01$), has been obtained from two independent datasets, namely a dense array of buoys in the Beaufort Sea during spring over a more limited scale range (8–140 km) (Hutchings et al. 2011), and another buoy array in the Weddell Sea (Antarctica) during austral spring (Hutchings et al. 2012).

The scaling law (3.2) holds over the entire spatial scale range available from RGPS, i.e. 10–1000 km, without detectable cut-off towards large scales. This is a signature of long-range spatial correlations, and of a correlation length larger than 1000 km (Girard et al. 2010), i.e. of the order of the size of the Arctic basin. Indeed, a random reshuffling of the strain-rate values destroys power law scaling (Marsan et al. 2004). As explain in Chaps. 4 and 5, this has strong consequences in terms of sea ice mechanical modeling. The analysis of the other moments reveals similar scaling laws (Marsan et al. 2004):

$$\langle \dot{\varepsilon}_{tot}^q \rangle \sim L^{-\beta(q)} \tag{3.3}$$

where $\beta(q)$ is a nonlinear function of q, expressing multifractality of sea ice deformation. As for velocity fluctuations in the time domain (see Sect. 2.4), strain-rates, i.e. velocity gradients, are characterized by a quadratic structure function, $\beta(q) = aq^2 + bq$ with $a = 0.13$ and $b = 0.07$ (Marsan et al. 2004), indicating that sea ice deformation can be modeled by a log-normal multiplicative cascade, a concept introduced initially for 3D fully developed turbulence (Yaglom 1966). The convexity of $\beta(q)$ ($a > 0$) implies that the higher moments of the PDF, corresponding to the most extreme strain-rates, grow faster towards small scales than for monofractal scaling (Marsan and Bean 2003). This means that the heterogeneity of sea ice deformation increases towards small scales, in full agreement with a decreasing power law exponent η (see above). Such multifractal analysis has not yet been performed independently on shear $\dot{\varepsilon}_{shear}$ and divergence $\dot{\varepsilon}_{div}$ strain invariants. However, owing to the strong correlations between $\dot{\varepsilon}_{tot}$, $\dot{\varepsilon}_{shear}$ and $\dot{\varepsilon}_{div}$ (Weiss and Schulson 2009), and to the power law PDFs $P(\dot{\varepsilon}) \sim \dot{\varepsilon}^{-\eta}$ obtained for both shear and divergence, one can reasonably assume that multifractal scaling holds for all components of sea ice deformation.

This multifractal scaling of sea ice deformation has several important consequences in terms of sea ice mechanics and modeling. The first, and most direct, is that the sea ice strain-rate is a scale dependent variable up to the correlation length, i.e. the scale of the Arctic basin as a whole. Consequently, great caution has to be taken when trying to use homogenization procedures to analyze/model sea ice deformation and rheology, and particularly to link different scales (Weiss and Marsan 2004). A striking example is given in Marsan et al. (2004). In material science and at the scale of laboratory experiments, strain rate is a key parameter controlling the physics of deformation, and therefore the rheology (Ashby 1972). For saline ice, a transition between ductile (dislocation related) and brittle rheology is observed under compressive loading around 10^{-4} s^{-1} (≈ 10 day^{-1})

(Schulson 2001; Schulson and Duval 2009), which represents the upper bound of the measured 10-km RGPS strain-rates (see Fig. 3.2). However, because of the scaling nature of the sea ice strain-rate fields, nothing can be directly deduced from these data about small-scale deformation mechanisms. The idea is therefore to use the multifractal properties to extrapolate the PDF of the total strain-rate to the 1 m scale, a scale similar to the laboratory scale as well as to the scale of sea ice thickness, a likely lower bound to scale invariance. This approach assumes that the scaling properties observed over the RGPS scale range (10–1000 km) extend down to ~ 1 m. As shown in Chap. 4, this hypothesis is supported by the strong analogies between deformation mechanisms observed at the lab and geophysical scales (Schulson 2004; Weiss 2003; Weiss and Schulson 2009). This rather involved extrapolation procedure, detailed in Marsan et al. (2004), shows that at such a small scale (and still at the 3-day timescale) about 15 % of the deformation, concentrated on only 0.2 % of the sea ice area, is accommodated in the brittle regime (>10 day^{-1}). As shown in Sect. 3.3, this localization of deformation is even reinforced towards smaller time scales as the result of intermittency, arguing for a sea ice deformation essentially accommodated by brittle failure over transient and very localized fracturing episodes. This is in qualitative agreement with the "old" hypothesis of a piecewise, almost-rigid motion of the sea ice cover (Nye 1975).

The scaling laws (3.2) and (3.3), which extend up to the 1000 km scale, show that there is no characteristic scale associated with sea ice deformation over the available scale range, in contradiction with some former assumptions (Overland et al. 1995). They reveal an associated correlation length of the order of the Arctic basin as a whole. This is the signature of long-range elastic interactions within the ice cover, allowing stresses and therefore brittle damage to be transmitted over very long distances (Girard et al. 2009). Consequently, these scaling properties can be proposed as validation metrics for sea ice mechanical models (Girard et al. 2009). It has been shown that the classical viscous-plastic framework (Hibler 1979) that is used nowadays in most climate models is inappropriate to represent the heterogeneity of sea ice deformation (Girard et al. 2009), as previously surmised by Nye (1973). Instead, an elasto-brittle framework, which considers sea ice as a continuous elastic plate encountering progressive damage simulating the initiation of cracks and leads, has been recently proposed and validated on this basis (Girard et al. 2011).

In a comprehensive analysis of the spatial scaling of sea ice deformation from RGPS data, (Stern and Lindsay 2009) showed that the scaling of sea ice deformation holds whatever the region of the Arctic, the ice category/age (first year vs. multiyear), or the season. However, they revealed a regional as well as seasonal dependence of the exponent $\beta(1)$, which increases during summer (Fig. 3.4) and/or in the peripheral zone of the Arctic or for first year sea ice. All of this argues for a larger exponent in the case of a looser, less cohesive sea ice cover. This is likely the signature of a transition in the mechanical behavior of sea ice, from an elasto-brittle plate with strong cohesion during winter in the central Arctic (Weiss and Schulson 2009; Weiss et al. 2007), to a more granular-like behavior in summer and/or in the marginal ice zone. In this last case, cohesion vanishes but the

3.2 Spatial Scaling and Localization of Deformation

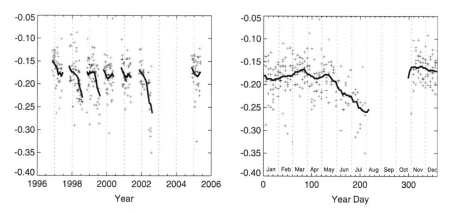

Fig. 3.4 Power-law spatial scaling exponent $-\beta(1)$ (*left*) by year and (*right*) by day of the year (all years from 1997 to 2005) for 297 snapshots (crosses). The bold curves on the left are 4-month running means and those on the right are 1-month running means. A clear increase (in absolute value) of the exponent is found during summer (from Stern and Lindsay 2009)

granular medium, in our case the assembly of floes, can still transmit compression as well as shear forces and stresses over long distances (Radjai et al. 1996; Taboada et al. 2005) and therefore allows scale invariance. Indeed, in the case of a frictional medium without cohesion, under multiaxial compression, a multifractal scaling (Eqs. 3.2 and 3.3) of both strain-rate and stress fields was obtained at the onset of flow instability, with $\beta(1) = 0.25 \pm 0.02$ for the shear strain-rate, and $\beta(1) = 0.38 \pm 0.02$ for the shear stress (Gimbert 2012). These exponents are larger than those reported for winter sea ice (0.18 ± 0.10), which are themselves in good agreement with the value obtained at failure for a cohesive elastic medium under similar loading, $\beta(1) = 0.15 \pm 0.02$ (Girard et al. 2010). The comparison of the sea ice scaling properties with the theoretical behavior of both elastic cohesive (Girard et al. 2010) and granular (Gimbert 2012) media, and particularly the fact that scaling holds over the entire scale range (10–1000 km), suggests that the sea ice cover remains permanently at the onset of mechanical instability.

This discussion shows that scaling properties can be an indicator of the mechanical behavior of sea ice and should guide the way towards new modeling approaches. Since an elasto-brittle modeling framework has been recently proposed and validated for the winter cohesive ice pack (Girard et al. 2011), granular-like discrete-element models could be an interesting alternative for a looser ice cover, such as in summer and/or in the marginal ice zone (Herman 2011; Hopkins et al. 2004; Wilchinsky et al. 2010).

Finally, it is worth noting that scale invariance of sea ice deformation can have a significant impact on the estimation of ice growth, mass balance, or ocean-atmosphere exchanges from kinematic data. Indeed, in winter, upward heat fluxes as well as new ice growth are directly related to divergence and lead opening. Consequently, the scaling of sea ice strain-rate implies that an estimation of these fluxes and ice production will be progressively underestimated as the resolution of the divergence data is improved (Hutchings et al. 2011). As an example, new ice production increases by a factor of about five for a spatial resolution going from 100 km to 10 km (Hutchings et al. 2011). Similar problems would arise for the estimation of closing rates and therefore ridge formation, or for downward heat fluxes in spring and summer. For these reasons, it would be important to know the lower bound to spatial scale invariance, which is so far unresolved by available RGPS or drifter data.

3.3 Spatial and Temporal Scaling Laws from the Dispersion of Lagrangian Trajectories

In the preceding section, we detailed the spatial scale invariance of sea ice deformation, but a possible dependence upon timescale was ignored. In Sect. 2.4, we showed the multifractality of velocity fluctuations in the time domain for individual Lagrangian trajectories, the signature of intermittency of sea ice drift. From this, we can conjecture, and actually expect, intermittency for sea ice deformation as well. As already noted in Sect. 3.1, the time resolution of RGPS is too coarse to properly perform such analysis. Drifter trajectories are recorded at a much higher frequency but are hampered by the scarcity of simultaneous records, except for some dense buoy arrays. For this reason, starting from the original work of Martin and Thorndike (Martin and Thorndike 1985), the idea is to analyze the dispersion of pairs of trajectories through time t and as a function of their initial separation $L(t = 0)$.

"Like two particles in a gas or two people in a crowd, two nearby pieces of sea ice gradually move apart and disperse" (Martin and Thorndike 1985). Thus, to study the dispersion of sea ice is a way to analyze its deformation (Thorndike 1986). The seminal analysis of these authors revealed, in a statistical sense, similarities between the dispersion properties of sea ice and the dispersion of particles in turbulent flow, although the underlying physics is different: in a dense ice pack, pieces of ice disperse as the result of fracture initiation, lead opening, and shear faulting, whereas eddies are at the origin of turbulent dispersion. The dispersion process in turbulent fluids, which has obvious consequences for e.g. the transport of pollutants (Pasquill 1974), has been studied for a long time (Batchelor 1952; Richardson 1926; Richardson and Stommel 1948) and, for geophysical fluids, experimentally tracked by either balloons in the atmosphere (Morel and Larchevèque 1974) or buoys at the surface of the ocean (Okubo 1971; Zhang et al. 2001). Note that the transport of pollutants by sea ice is also of practical importance (Pavlov et al. 2004), but will not be addressed here in detail.

As noted earlier, strain tensors cannot be obtained from pairs of trajectories. Therefore, we aim at defining a proxy of strain-rate that will depend on divergence, convergence and shear, but will be insensitive to solid rotation. Inspired by the classical methodology developed in turbulence (Batchelor 1952; Richardson and

3.3 Spatial and Temporal Scaling Laws from the Dispersion

Stommel 1948), Martin and Thorndike (Martin and Thorndike 1985) and more recently Rampal et al. (2008) considered pairs of ice drifters, noted 1 and 2 with positions $\mathbf{X_1}(t)$ and $\mathbf{X_2}(t)$ and separation $\mathbf{Y}(t) = \mathbf{X_1}(t) - \mathbf{X_2}(t)$ (Fig. 3.5). If the two tracers, initially separated by a spatial scale $L = Y(0)$, are observed after a time τ, their separation changes to $l(\tau) = Y(\tau)$. The change in separation is then defined as:

$$\Delta r = Y(\tau) - Y(0) = l(\tau) - L \tag{3.4}$$

If one aims at comparing dispersion regimes between fluid turbulence and sea ice, the mean squared change of separation $\langle \Delta r^2 \rangle$ can be analyzed (see below). However, from a solid mechanics perspective, dealing with a deformation rate seems more pertinent. Consequently, one can define a proxy of strain-rate based on the standard deviation of the dispersion normalized by length and time:

$$\dot{\varepsilon}_{disp} = \frac{\left(\langle \Delta r^2 \rangle - \langle \Delta r \rangle^2\right)^{1/2}}{L\tau} \tag{3.5}$$

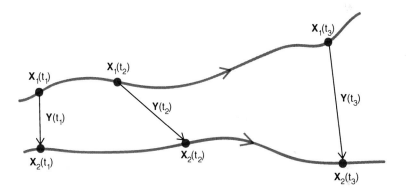

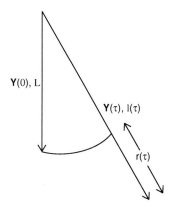

Fig. 3.5 For a pair of buoys located at $\mathbf{X_1}(t)$, $\mathbf{X_2}(t)$, the separation $\mathbf{Y}(t) = \mathbf{X_2}(t) - \mathbf{X_1}(t)$ yields the dispersion $\Delta r = Y(\tau) - Y(0)$ (from Rampal et al. 2008)

The first step is to validate this definition of a strain-rate proxy from a dataset consisting of a dense array of buoys, allowing a simultaneous estimate of the proxy $\dot{\varepsilon}_{disp}$ as well as of the classical total strain-rate $\dot{\varepsilon}_{tot}$. This was done from the SHEBA array (Uttal et al. 2002) by Rampal et al. (2008) who showed that in this particular case $\dot{\varepsilon}_{disp} \approx 2\dot{\varepsilon}_{tot}$ with a correlation coefficient larger than 0.9 (Fig. 3.6). The dispersion analysis can be extended to the full IABP dataset to study the scaling properties of sea ice deformation simultaneously in the time and space domains (Rampal et al. 2008). To analyze spatial scaling, pairs of drifters were grouped in terms of initial separation L. A spatial scaling,

$$\langle \dot{\varepsilon}_{disp} \rangle \sim L^{-\beta(\tau)} \tag{3.6}$$

was observed (Fig. 3.7b), in full qualitative agreement with the RGPS analysis and Eq. (3.2). A full multifractal analysis was not performed in this case. Instead, a dependence of the spatial scaling exponent β upon timescale τ was observed: β increases with decreasing timescale, from $\beta \approx 0.35$ for $\tau = 2$ months (in winter) to $\beta \approx 0.85$ for $\tau = 3$ h (Rampal et al. 2008) (Fig. 3.7b and 3.8b). No significant differences were observed between winter and summer scaling, but we should stress here that the uncertainty in the estimation of the exponents is larger

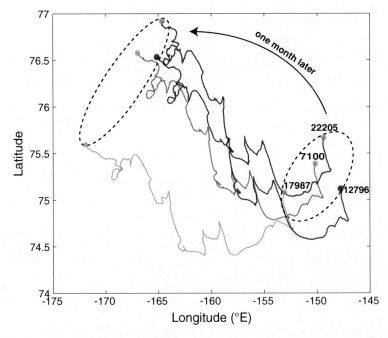

Fig. 3.6 The dispersion of Lagrangian tracers (buoys) as a measure of sea ice deformation: the case of the SHEBA experiment. The tracks of 4 SHEBA buoys are shown from October 27th to November 24th, 1997. The evolution of the smallest ellipse including all buoys illustrates the deformation of sea ice with a combination of shear and divergence (adapted from Rampal et al. 2008)

3.3 Spatial and Temporal Scaling Laws from the Dispersion

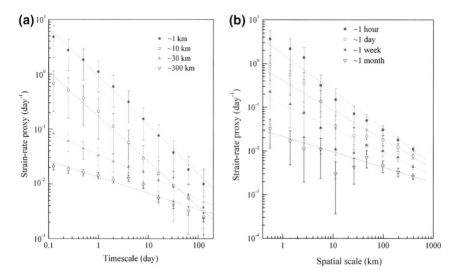

Fig. 3.7 Temporal (*left*) as well as spatial (*right*) scaling of the strain-rate proxy $\dot{\varepsilon}_{disp}$ in winter calculated from the IABP data (1979–2002). The temporal (respectively spatial) scaling depends on the spatial (respectively time) scale considered, thus illustrating the space/time coupling of Arctic sea ice deformation. The errors on the estimation of $\dot{\varepsilon}_{disp}$ have been calculated from a bootstrap method (Rampal et al. 2008)

for this dataset, ranging from 15 to 80 %. The corresponding β-value for a 3-day (RGPS) timescale is about 0.6, i.e. substantially larger than the RGPS exponent, $\beta \approx 0.2$. This discrepancy can be only partly explained by the large uncertainty on the buoy-derived exponent, and therefore suggests a more complex, non-linear relation between the proxy $\dot{\varepsilon}_{disp}$ and $\dot{\varepsilon}_{tot}$. This point would deserve further attention.

Similar to the spatial scaling, a temporal scaling law emerged:

$$\langle \dot{\varepsilon}_{disp} \rangle \sim \tau^{-\alpha(L)} \tag{3.7}$$

with an exponent α increasing with decreasing spatial scale L, from $\alpha \approx 0.3$ for $L \sim 300$ km to $\alpha \approx 0.9$ for $L \sim 1$ km (Figs. 3.7a and 3.8a). Once again, no significant differences were observed between winter and summer, although the uncertainties on α were smaller, ranging from 7 to 25 %. A similar behavior has been observed for strain-rates obtained from a dense array of buoys in the Beaufort Sea (Hutchings et al. 2011).

As detailed in next section, these two scaling laws (3.6) and (3.7) are actually the two symmetric faces of a unified space/time coupling symmetry (Marsan and Weiss 2010). It will be shown in Chap. 4 that this space/time coupling is the fingerprint of an underlying brittle deformation process accommodated by discrete fracturing events at various scales—a deformation mechanism already suggested by several analyses detailed above.

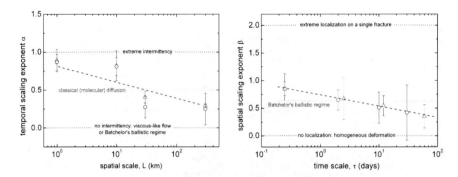

Fig. 3.8 Space-time coupling in Arctic sea ice deformation. **a** Evolution of the temporal scaling exponent α, characterizing the intermittency of deformation, with the spatial scale L (see Fig. 3.7). *Triangles*: winter; *circles*: summer. The *dashed line* represents a logarithmic scaling $\alpha \sim c \times \ln(L)$ with $c = 0.10$. The α-values are bounded by $\alpha = 0$ corresponding to no intermittency (or Batchelor's ballistic regime Batchelor 1950), and by $\alpha = 1$ corresponding to extreme intermittency, i.e. a deformation process due to a unique event in time. **b** Evolution of the spatial scaling exponent β, characterizing the spatial heterogeneity of deformation, with the timescale τ (see Fig. 3.7). *Triangles*: winter; *circles*: summer. The *dashed line* represents a logarithmic scaling $\beta \sim c \times \ln(\tau)$ with $c = 0.10$. The β-values are bounded by $\beta = 0$ corresponding to perfectly homogeneous deformation, and by $\beta = 2$ corresponding to extreme localization at a single "point". The errors on the estimation of the exponents are related to the errors on the estimation of the strain-rate proxy $\dot{\varepsilon}_{disp}$ (see Fig. 3.7 and Rampal et al. 2008). (adapted from Marsan and Weiss 2010)

3.4 Space/Time Coupling

As discussed in Sect. 3.5, the coupled Eqs. (3.6) and (3.7) cannot be explained by the turbulent properties of the forcing fields (atmospheric or oceanic turbulence). Instead, an intrinsic origin resulting from the mechanical behavior of the sea ice cover has to be sought. The scaling exponents α and β can be interpreted as follows:

- The exponent β expresses the degree of spatial heterogeneity/localization of sea ice deformation, bounded by $\beta = 0$ for a perfectly homogeneous deformation field (e.g. viscous-like), and by $\beta = 2$, i.e. the topological dimension for the 2D-like sea ice cover, for a deformation localized on a single "point" (Marsan and Weiss 2010; Rampal et al. 2008). In that sense, sea ice deformation appears slightly more localized in summer (see Sect. 3.2 and Stern and Lindsay 2009).
- Similarly, α is a measure of the degree of intermittency of the deformation process, bounded by $\alpha = 0$ for a non-intermittent viscous-like flow, and by $\alpha = 1$ for a deformation accommodated by a unique "event".

The decrease of β upon increasing the timescale of observation means that sea ice deformation appears smoother, more homogeneous towards large timescales, but localized deformation features persist over a season as β remains strictly positive, in agreement with an analysis of Coon et al. (2007). Similarly, the

3.4 Space/Time Coupling

decrease of α as one coarsens the spatial scale means that intermittency decreases accordingly, but nevertheless persists for $L = 300$ km ($\alpha > 0$), i.e. sea ice deformation does not mimic viscous-like flow even at scales comparable to the size of the arctic basin (Marsan and Weiss 2010).

The dependence of exponent α (respectively β) on spatial scale (respectively time scale) implies that $\dot{\varepsilon}(L, \tau) \sim \tau^{-\alpha(L)} \sim L^{-\beta(\tau)}$. The only compatible expression is the coupled equation (Marsan and Weiss 2010):

$$\dot{\varepsilon}(L, \tau) \sim \tau^{-\alpha_0} L^{-\beta_0} e^{c \ln(\tau) \ln(L)} \tag{3.8}$$

which thus gives that $\dot{\varepsilon}(L) \sim L^{(-\beta_0 + c\ln(\tau))}$ hence $\beta(\tau) = \beta_0 - c\ln(\tau)$, and $\dot{\varepsilon}(\tau) \sim \tau^{(-\alpha_0 + c\ln(L))}$ hence $\alpha(L) = \alpha_0 - c\ln(L)$. Parameter c represents the strength of the space/time coupling, whereas α_0 and β_0 are constants which values depend on the chosen space and time units, and thus that have no particular physical meaning. The observed dependence of α on τ and β on L for Arctic sea ice (Rampal et al. 2008) are in reasonable agreement with this expression, with $c \approx 0.10$ (Fig. 3.8).

As detailed in next chapter, a similar space/time scaling symmetry is observed for the seismicity of the Earth's crust, reinforcing the analogy between the brittle deformation of the sea ice cover and of the upper crust. In Sect. 3.2, we have seen that an extrapolation of the PDF of the total strain-rate $\dot{\varepsilon}_{tot}$ to the 1 m scale, however still at the RGPS time resolution of 3 days, showed that at these scales about 15 % of the deformation was accommodated in a purely brittle regime (>10 day^{-1}). The results of Fig. 3.7b show that intermittency keeps increasing below $\tau = 3$ days, which should reinforces this point. A coupled space/time scaling analysis has not yet been performed on the full strain-rate PDF, so a similar extrapolation cannot be performed for timescales below 3 days. However, we can already notice that the average total strain-rate $\langle \dot{\varepsilon}_{tot} \rangle \approx \frac{1}{2} \langle \dot{\varepsilon}_{disp} \rangle$ that can be estimated from buoy dispersion analysis at the smallest available scales (3 h, ~ 1 km) is $\langle \dot{\varepsilon}_{tot} \rangle \approx 2.5$ day^{-1}, a value not so far from the ductile-to-brittle transition strain rate of ice in laboratory experiments ($\dot{\varepsilon}_{D/B} \approx 10^{-4}$ s$^{-1} \approx 10$ day^{-1}). A simple spatial downscaling can easily reduce the remaining gap. Indeed, using $\beta = 0.85$ obtained for $\tau = 3$ h, one finds $\langle \dot{\varepsilon}_{tot} \rangle \approx 10$ day^{-1} for a length scale $L \approx \left(\frac{10 \, day^{-1}}{2.5 \, day^{-1}} \right)^{-\frac{1}{0.85}} \approx 0.2$ km. Using a much smaller exponent $\beta = 0.2$, on still find $\langle \dot{\varepsilon}_{tot} \rangle \approx 10$ day^{-1} for a length scale of 1 m. In addition, owing to the strongly non-gaussian, power law character of strain-rate PDFs (see Sect. 3.2), one expects that at these small space and time scales nearly all the deformation is accommodated by brittle processes. Note that an elastic wave propagating at a speed of ~ 400 m/s (Marsan et al. 2011) would be able to travel across the entire arctic basin in much less than 3 h, indicating the relevance of such small timescales for brittle processes. Here, we did not considered the possible presence of defects, such as pre-existing thermal cracks, within the ice cover. As argued by Schulson (1990, 2001) and briefly explained in Sect. 4.3, the presence of such "stress concentrators" decreases the transition strain-rate, making the brittle character of the sea ice cover even more plausible.

3.5 Sea Ice Dispersion as the Result of "Solid Turbulence"

Although the nature of the "brittle" dispersion process detailed in Sect. 3.3 is fundamentally different from the processes involved in fluid dispersion (molecular dispersion and turbulence), it is interesting to compare the above space and time scaling laws with corresponding dispersion laws of fluid mechanics. Recast in terms of pair dispersion, Eqs. (3.6) and (3.7) respectively read

$$\langle \Delta r^2 \rangle - \langle \Delta r \rangle^2 \sim L^\delta = L^{2(1-\beta)} \tag{3.9a}$$

and

$$\langle \Delta r^2 \rangle - \langle \Delta r \rangle^2 \sim \tau^\gamma = \tau^{2(1-\alpha)} \tag{3.9b}$$

These scaling laws can be compared to theoretical predictions for homogeneous fluid turbulence (see e.g. Sawford 2001 for a review). Within the small-scale dissipation range, i.e. for an initial separation below the Kolmogorov length scale, $L \ll L_\eta$, and timescales well below the dissipative timescale, $\tau \ll t_\eta$, the velocity field is reasonably assumed to vary smoothly and one expects an exponential growth of the relative dispersion, $\langle \Delta r^2 \rangle \sim L^2 \exp\left(\frac{\tau}{c_1 t_\eta}\right)$, where c_1 is a constant (Monin and Yaglom 1975). Within the inertial range, i.e. $t_\eta \ll \tau \ll \Gamma$, two super-diffusive regimes ($\gamma > 1$) are predicted (Batchelor 1950), instead of one in the case of the diffusion of individual trajectories (see Sect. 2.3). For times below a characteristic timescale t_0 that depends on the initial separation L as $t_0 \sim L^{2/3}$, a ballistic regime is characterized by $\langle \Delta r^2 \rangle \sim L^{2/3} \tau^2$, whereas Richardson's regime (Richardson 1926) $\langle \Delta r^2 \rangle \sim \tau^3$ is expected for $\tau \gg t_0$, with no dependence upon the initial separation ($\delta = 0$). For timescales well above the integral time of turbulence, $\tau \gg \Gamma$, a classical diffusive regime ($\gamma = 1$) is recovered. It is worth noting that the confirmation of these theoretical predictions is still controversial. Laboratory experiments agreed with Batchelor's $L^{2/3} \tau^2$ scaling, but did not find evidence of Richardon's τ^3 regime (Bourgoin et al. 2006). For geophysical fluids, the situation is even less clear: for atmospheric turbulence, the same dataset (the EOLE experiment on the dispersion of balloons) has been either interpreted in terms of exponential growth (Morel and Larchevèque 1974) or of τ^3 scaling (Lacorata et al. 2004). For dispersion in the ocean, several studies supported Richardson's scaling (e.g. Okubo 1971; Ollitrault et al. 2005), some showed a transition from exponential growth to Richardson's scaling around a spatial scale of a few tens of km (Ollitrault et al. 2005), whereas others argued for γ-values well below three and/or a regional dependence of the dispersion regime (for a recent review see e.g. LaCasce 2008).

3.5 Sea Ice Dispersion as the Result of "Solid Turbulence"

In any case, the dispersion of sea ice departs significantly from turbulent dispersion in several ways:

(i) The space and time scaling analyses performed by Rampal et al. (2008) did not reveal any characteristic scale, but instead a continuous evolution of both the temporal (respectively spatial) scaling upon the spatial (respectively temporal) scale considered. Eq. (3.7) and Fig. 3.7b show that some correlation and memory persist in the deformation process up to the timescale of a few months. As already stressed above, this is the fingerprint of a space/time symmetry which cannot be explained by turbulence theory.
(ii) In particular, no transition in terms of dispersion or strain-rate scaling is observed across the integral timescale of ~ 1.5 days obtained in Sect. 2.2.
(iii) Recast in terms of relative dispersion Eq. (3.9a, b), the temporal scaling exponents α of sea ice strain rate give γ-values varying from a slightly super-diffusive regime at small spatial scale ($\gamma = 1.4$ for $L = 1$ km) to a very slow sub-diffusive regime at large scales ($\gamma = 0.2$ for $L = 300$ km). This means that sea ice dispersion is much slower than oceanic dispersion. Important consequences are expected in terms of pollutant dispersion and transport, a problem that may increase in the future with growing industrial activities in the Arctic (see e.g. Pavlov et al. 2004; Pfirman et al. 1997 about transport of contaminants by sea ice).

3.6 Spectral Analysis

In Sect. 3.3, intermittency of sea ice deformation was analyzed in the time domain. It can alternatively be explored from a spectral analysis. Indeed, Eq. (3.7) implies that the average strain increment measured over a time window τ scales as $\langle \varepsilon \rangle \sim \tau^{1-\alpha(L)}$. Considering this strain record as a random walk, this scaling law would correspond to a self-affine profile with a "roughness" exponent $H = 1 - \alpha$. Consequently, the corresponding power spectrum would scale as $p_\varepsilon(\omega) \sim \omega^{-(2H+1)} = \omega^{-3+2\alpha}$ (Schmittbuhl et al. 1995), and therefore the power spectrum of the strain-rate $\dot{\varepsilon}$ should scale as $p_{\dot{\varepsilon}}(\omega) \sim \omega^{-(2H-1)} = \omega^{-1+2\alpha}$. From the scaling analysis in the time domain (Rampal et al. 2008), one thus expects for $\alpha > 0.5$, i.e. at small spatial scales ($L \leq 10$ km, see Fig. 3.8a), a strain-rate power spectrum increasing towards higher frequencies. This strong intermittency corresponds to an anti-persistent record (Mandelbrot and Van Ness, 1968) for which a brief and intense strain-rate event is expected, in the probabilistic sense, to be preceded and/or followed by quiescent periods. On the opposite, at larger spatial scales ($L > 10$ km), $\alpha < 0.5$ and consequently $p_{\dot{\varepsilon}}(\omega)$ decreases towards high frequencies, a signature of a persistent signal with positive correlations of the strain-rate record. Note that the value $\alpha = 0.5$ would correspond to classical random walk without any correlation between successive strain-rates, i.e. a molecular-like dispersion process (Fig. 3.8a).

Such spectral analysis has not been performed systematically at the scale of the Arctic basin. From their buoy array in the Beaufort sea, (Hutchings et al. 2011) obtained over a limited frequency range (0.2–6 day^{-1}) power spectra that are compatible with power law scaling $p_{\dot{\varepsilon}}(\omega) \sim \omega^{-\varsigma}$, with an exponent ς that increases as one coarsens the spatial scale of observation, in agreement with the former expectation. However, this exponent was always found to be positive, i.e. the strain-rate record to be persistent, even at the scale of 10 km.

It is interesting to note that, as for the sea ice velocity spectrum (see Sect. 2.6), the strain-rate power spectrum can sometimes be modulated by a semi-diurnal oscillation (Heil et al. 2008; Kwok et al. 2003). This can be understood as a differential effect of the Coriolis forcing across a given zone as the result of spatial variations in sea ice concentration, thickness or strength. This effect seems particularly marked for divergence, implying recurrent openings and closing that may have an impact on new ice production during winter (Kwok et al. 2003).

3.7 Concluding Remarks

In Chap. 2, sea ice drift was explored by means of the analysis of individual Lagrangian trajectories. Strong analogies with fluid turbulence were stressed, and they argued for an inheritance from atmospheric and oceanic forcing, although the intermittency of sea ice motion appeared more pronounced, suggesting a non-linear rheological response of sea ice. In the present chapter, sea ice deformation, or alternatively the relative dispersion of tracers, was investigated. Although this analysis was partly performed from statistical tools previously developed for turbulence, strong differences between sea ice and turbulent fluids have been revealed. In particular, a strong space/time coupling characterizes sea ice deformation, meaning that the associated intermittency and the spatial localization increase towards respectively small spatial (or temporal) scales. Such space/time scaling symmetry has no equivalence in fluid turbulence but, as shown in the next chapter, has been documented for the brittle deformation of the upper Earth's crust. This argues for a strongly non-linear, brittle rheology of the sea ice cover as well, a mechanical behavior further supported by an extrapolation of sea ice strain-rate PDFs towards small space and time scales and a comparison with the ductile-to-brittle transition strain-rate obtained from laboratory mechanical tests.

Still, to analyze rheology, a measure of in situ stress states is needed. The characteristics of such internal stress records are summarized in the next chapter, along with an analysis of the scaling properties of sea ice fracture and fragmentation patterns that confirm the brittle behavior of sea ice spanning an extremely large range of spatial and temporal scales.

References

Ashby, M. F. (1972). A first report on deformation-mechanism maps. *Acta Metallurgica, 20*, 887–897.

Batchelor, G. K. (1950). The application of the similarity theory of turbulence to atmospheric diffusion. *Quarterly Journal of the Royal Meteeorological Society, 76*, 133–146.

Batchelor, G. K. (1952). Diffusion in a field of homogeneous turbulence. II. The relative motion of particles. *Proceedings of the Cambridge Philological Society, 48*, 345–362.

Bourgoin, M., Ouellette, N. T., Xu, H. T., Berg, J., & Bodenschatz, E. (2006). The role of pair dispersion in turbulent flow. *Science, 311*(5762), 835–838.

Coon, M., Kwok, R., Levy, G., Pruis, M., Schreyer, H., & Sulsky, D. (2007). Arctic ice dynamics joint experiment (AIDJEX) assumptions revisited and found inadequate. *Journal of Geophysical Research, 112*, C11S90.

Gimbert, F. (2012). *Mechanics of Arctic sea ice and granular materials: two media, two approaches*. PhD Thesis, University of Grenoble, Grenoble.

Girard, L., Amitrano, D, & Weiss, J. (2010). Fracture as a critical phenomenon in a progressive damage model. *Journal of Statistical Mechanics*, P01013.

Girard, L., Bouillon, S., Weiss, J., Amitrano, D., Fichefet, T., & Legat, V. (2011). A new modelling framework for sea-ice mechanics based on elasto-brittle rheology. *Annals of Glaciology, 52*(57), 123–132.

Girard, L., Weiss, J., Molines, J. M., Barnier, B., & Boullion, S. (2009). Evaluation of two sea ice models on the basis of statistical and scaling properties of Arctic sea ice deformation. *Journal of Geophysical Research, 114*, C08015.

Heil, P., Hutchings, J. K., Worby, A. P., Johansson, M., Launiainen, J., Haas, C., et al. (2008). Tidal forcing on sea-ice drift and deformation in the western Weddell Sea in early austral summer, 2004. *Deep-Sea Research Part II, Topical Studies in Oceanography, 55*(8–9), 943–962.

Herman, A. (2011). Molecular-dynamics simulation of clustering processes in sea-ice floes. *Physical Review E, 84*(5), 056104 .

Hibler, W. D. I. (1979). A dynamic thermodynamics sea ice model. *Journal of Physical Oceanography, 9*, 815–846.

Hopkins, M. A., Frankenstein, S., & Thorndike, A. S. (2004). Formation of an aggregate scale in Arctic sea ice. *Journal of Geophysical Research, 109*, C01032.

Hutchings, J. K., Heil, P., Steer, A., & Hibler, W. D. (2012). Subsynoptic scale spatial variability of sea ice deformation in the western Weddell Sea during early summer. *Journal of Geophysical Research-Oceans, 117*, C01002.

Hutchings, J. K., Roberts, A., Geiger, C. A., & Richter-Menge, J. A. (2011). Spatial and temporal characterization of sea ice deformation. *Annals of Glaciology, 52*(57), 360–368.

Jaeger, J. C., & Cook, N. G. W. (1979). *Fundamentals of rock mechanics*. London: Chapman and Hall.

Kwok, R. (1998). The RADARSAT geophysical processor system. In C. Tsatsoulis & R. Kwok (Eds.), *Analysis of SAR data of the polar oceans* (pp. 235–257). Springer-Verlag.

Kwok, R. (2001). *Deformation of the arctic ocean sea ice cover between November 1996 and April 1997: a survey, paper presented at IUTAM scaling laws in ice mechanics and ice dynamics*. Fairbanks: Kluwer Academic Publishers.

Kwok, R., Cunningham, G. F., & Hibler, W. D. I. (2003). Sub-daily sea ice motion and deformation from RADARSAT observations. *Geophysical Research Letters, 30*(23), L018723.

Kwok, R., Curlander, J. C., McConnell, R., & Pang, S. S. (1990). An ice-motion tracking system at the Alaska SAR facility. *IEEE Journal of Oceanic Engineering, 15*(1), 44–54.

LaCasce, J. H. (2008). Statistics from Lagrangian observations. *Progress in Oceanography, 77*(1), 1–29.

Lacorata, G., Aurell, E., Legras, B., & Vulpiani, A. (2004). Evidence for a $k^{-5/3}$ spectrum from the EOLE Lagrangian balloons in the low stratosphere. *Journal of Atmospheric Sciences, 61*, 2936–2942.

Lindsay, R. W., & Stern, H. L. (2003). The RadarSat geophysical processor system: quality of sea ice trajectory and deformation estimates. *Journal of Atmospheric and Oceanic Technology, 20*, 1333–1347.

Mandelbrot, B., & Van Ness, J. (1968). Fractional Brownian motions, fractional noises and applications. *SIAM Review, 10*(4), 422–437.

Marsan, D., & Bean, C. J. (2003). Multifractal modeling and analyses of crustal heterogeneity. In J. A. Goff & K. Holliger (Eds.), *Heterogeneity in the Crust and Upper Mantle* (pp. 207–236). Kluwer Academic Publishers.

Marsan, D., Stern, H., Lindsay, R., & Weiss, J. (2004). Scale dependence and localization of the deformation of arctic sea ice. *Physical Review Letters, 93*(17), 178501.

Marsan, D., & Weiss, J. (2010). Space/time coupling in brittle deformation at geophysical scales. *Earth and Planetary Science Letters, 296*(3–4), 353–359.

Marsan, D., Weiss, J., Metaxian, J. P., Grangeon, J., Roux, P. F., & Haapala, J. (2011). Low-frequency bursts of horizontally polarized waves in the Arctic sea-ice cover. *Journal of Glaciology, 57*(202), 231–237.

Martin, S., & Thorndike, A. S. (1985). Dispersion of sea ice in the Bering Sea. *Journal of Geophysical Research, 90*(C4), 7223–7226.

Monin, A. S., & Yaglom, A. M. (1975). *Statistical fluid mechanics: Mechanics of Turbulence* (p. 874). Cambridge: MIT Press.

Morel, P., & Larchevêque, M. (1974). Relative dispersion of constant-level balloons in the 200-mb general circulation. *Journal of Atmospheric Science, 31*, 2189–2196.

Moritz, R. E., & Stern, H. L. (2001). Relationships between geostrophic winds, ice strain rates and the piecewise rigid motions of pack ice. In J. P. Dempsey & H. H. Shen (Eds.), *Scaling laws in ice mechanics and ice dynamics* (pp. 335–348). Dordrecht: Kluwer Academic Publishers.

Nye, J. F. (1973). Is there any physical basis for assuming linear viscous behavior for sea ice. *AIDJEX Bulletin, 21*, 18–19.

Nye, J. F. (1975). The use of ERTS photographs to measure the movement and deformation of sea ice. *Journal of Glaciology, 15*(73), 429–436.

Okubo, A. (1971). Oceanic diffusion diagrams. *Deep-Sea Research, 18*, 789–802.

Ollitrault, M., Gabillet, C., & De Verdiere, A. C. (2005). Open ocean regimes of relative dispersion. *Journal of Fluid Mechanics, 533*, 381–407.

Overland, J. E., Walter, B. A., Curtin, T. B., & Turet, P. (1995). Hierarchy and sea ice mechanics: A case study from the Beaufort Sea. *Journal of Geophysical Research, 100*(C3), 4559–4571.

Pasquill, F. (1974). *Atmospheric diffusion* (2nd ed.). New York: Wiley.

Pavlov, V., Pavlova, O., & Korsnes, R. (2004). Sea ice fluxes and drift trajectories from potential pollution sources, computed with a statistical sea ice model of the Arctic Ocean. *Journal of Marine Systems, 48*(1–4), 133–157.

Pfirman, S. L., Kogeler, J. W., & Rigor, I. (1997). Potential for rapid transport of contaminants from the Kara Sea. *Science of the Total Environment, 202*(1–3), 111–122.

Radjai, F., Jean, M., Moreau, J. J., & Roux, S. (1996). Force distributions in dense two-dimensional granular systems. *Physical Review Letters, 77*(2), 274–277.

Rampal, P., Weiss, J., Marsan, D., Lindsay, R., & Stern, H. (2008). Scaling properties of sea ice deformation from buoy dispersion analyses. *Journal of Geophysical Research, 113*, C03002.

Richardson, L. F. (1926). Atmospheric diffusion shown on a distance-neighbour graph. *Proceedings of the Royal Society of London A, 110*, 709–737.

Richardson, L. F., & Stommel, H. (1948). Note on eddy diffusion in the sea. *Journal of Meteorology, 5*, 238–240.

Sawford, B. (2001). Turbulent relative dispersion. *Annual Review of Fluid Mechanics, 33*, 289–317.

References

Schmittbuhl, J., Schmitt, F., & Scholz, C. (1995). Scaling invariance of crack surfaces. *Journal of Geophysical Research, 100*, 5953–5973.

Schulson, E. M. (1990). The brittle compressive fracture of ice. *Acta Metallurgica et Materialia, 38*(10), 1963–1976.

Schulson, E. M. (2001). Brittle failure of ice. *Engineering Fracture Mechanics, 68*(17–18), 1839–1887.

Schulson, E. M. (2004). Compressive shear faults within the arctic sea ice: Fracture on scales large and small. *Journal of Geophysical Research, 109*, C07016, C002108.

Schulson, E. M., & Duval, P. (2009). *Creep and fracture of ice.* Cambridge: Cambridge University Press.

Sornette, D. (2000). *Critical phenomena in natural sciences.* Berlin: Springer.

Stern, H. L., & Lindsay, R. W. (2009). Spatial scaling of Arctic sea ice deformation. *Journal of Geophysical Research, 114*, C10017.

Stern, H. L., Rothrock, D. A., & Kwok, R. (1995). Open water production in Arctic sea ice: Satellite measurements and model parametrizations. *Journal of Geophysical Research, 100*(C10), 20601–20612.

Taboada, A., Chang, K. J., Radjai, F. & Bouchette, F. (2005). Rheology, force transmission, and shear instabilities in frictional granular media from biaxial numerical tests using the contact dynamics method. *Journal of Geophysical Research-Solid Earth, 110*(B9).

Thomas, M., Geiger, C. A., & Kambhamettu, C. (2008). High resolution (400 m) motion characterization of sea ice using ERS-1 SAR imagery. *Cold Regions Science and Technology, 52*, 207–223.

Thorndike, A. S. (1986). Kinematics of sea ice. In N. Untersteiner (Ed.), *The geophysics of sea ice* (pp. 489–549). New York: Plenum Press.

Thorndike, A. S., & Colony, R. (1982). Sea ice motion in response to geostrophic winds. *Journal of Geophysical Research, 87*(C8), 5845-5852.

Thorndike, A. S., Rothrock, D. A., Maykut, G. A., & Colony, R. (1975). The thickness distribution of sea ice. *Journal of Geophysical Research, 80*(33), 4501–4513.

Uttal, T., et al. (2002). Surface heat budget of the Arctic Ocean. *Bulletin of the American Meteorological Society, 83*(2), 255–275.

Weiss, J. (2003). Scaling of fracture and faulting in ice on Earth. *Surveys of Geophysics, 24*, 185–227.

Weiss, J., & Marsan, D. (2004). Scale properties of sea ice deformation and fracturing. ***Comptes Rendus Physique***, *5*(7), 683–685.

Weiss, J., & Schulson, E. M. (2009). Coulombic faulting from the grain scale to the geophysical scale: Lessons from ice. *Journal of Physics. D. Applied Physics, 42*, 214017.

Weiss, J., Schulson, E. M., & Stern, H. L. (2007). Sea ice rheology from in situ, satellite and laboratory observations: Fracture and friction. *Earth and Planetary Science Letters, 255*, 1–8.

Wilchinsky, A. V., Feltham, D. L., & Hopkins, M. A. (2010). Effect of shear rupture on aggregate scale formation in sea ice. *Journal of Geophysical Research-Oceans, 115*.

Yaglom, A. M. (1966). The Influence of Fluctuations in Energy Dissipation in the Shape of Turbulence Characteristics in the Inertial Interval. *Soviet Physics. Doklady, 2*, 26–30.

Zhang, H. M., Prater, M. D., & Rossby, T. (2001). Isopycnal lagrangian statistics from the North Atlantic current RAFOS float observations. *Journal of Geophysical Research, 106*(C7), 13817–13836.

Chapter 4
Sea Ice Fracturing

Abstract Two very different types of data are reviewed in this chapter: in-situ internal stress measurements in the one hand, and fracture, fault and fragmentation patterns as seen in aerial or satellite images on the other hand. In-situ stress records show that sea ice internal stress states lie within a Coulombic envelope whose shape and the associated internal friction coefficient are very similar to that obtained from brittle compressive failure tests on saline ice in the laboratory. This argues for a brittle rheology of sea ice involving faulting and friction. However, the sea ice shear strength, which can be unambiguously estimated around 30 kPa, is significantly lower than laboratory measurements, thus implying a scale effect on strength. In-situ stress records are also highly intermittent, either in terms of intensity or principal stress directions. This intermittency cannot be explained by wind forcing characteristics, but is instead the signature of short and transient fracturing/faulting episodes. In the spatial domain, these fracturing events generate scale invariant fracture networks. From these observations, a multiscale statistical model based on the superposition of numerous discrete displacements associated to fracturing/faulting episodes can be built to explain the characteristics of sea ice deformation.

Fracture and fragmentation of the sea ice cover are ubiquitous in all seasons, empirically known for thousands of years by the people whose subsistence depended on sea ice through hunting and fishing and, as shown in this chapter, encountered at all scales, up to the scale of the Arctic basin as a whole. These processes are so ubiquitous and important that they are associated with specific terminology: sea ice *leads* for fractures and shear faults, ice *floes* for the pieces of ice resulting from the fragmentation of the ice cover. While they are obstacles for skiers or sledges, sea ice leads, which can extend over hundreds of kilometers, can sometimes provide opportunities for easier traveling for ships and icebreakers. In what follows, we use the term "fracture" in a generic sense, whereas "fault" more specifically refers to fracturing and relative motion across a discontinuity under shear.

Nowadays, in the context of Arctic amplification of climate change (Serreze and Francis 2006) associated with a shrinking ice cover (Comiso et al. 2008;

Lemke and others 2007), the role of opening leads in ocean-atmosphere interactions is fundamental, although they cover at most 1–2 % in winter and 5–12 % in summer of the total sea ice surface (Fichelet and Morales Maqueda 1995). Through leads, air and water come into contact and directly exchange latent and sensible heat through convective processes driven by the large temperature difference between them (up to 30–40 °C in winter). In summer, leads also play a large role in the absorption of shortwave radiation due to the low albedo of open water (<0.2), compared to the albedo of multiyear sea ice (>0.7) (Fichelet and Morales Maqueda 1995). The upward heat fluxes between open water and the atmosphere are orders of magnitude larger than through thick ice: turbulent heat fluxes (sensible and latent) are less than 5 $W \cdot m^{-2}$ over multiyear ice (Maykut 1982) and can be up to 600 $W \cdot m^{-2}$ over open water (Andreas and Murphy 1986; Maykut 1986). Consequently, the variability of sea ice fracturing could have a large impact on climate: as an example, for a sea ice cover with an open water fraction as little as 0.5 %, this fraction will contribute about half of the total thermal energy transfer between the ocean and the atmosphere (Heil and Hibler 2002).

The role of fracturing processes in sea ice deformation has been recognized for a long time (e.g. Nye 1973, 1975), but analyzed in detail only recently (Marsan and Weiss 2010; Schulson 2004; Schulson and Hibler 1991, 2004; Weiss et al. 2007, 2009). In previous chapters, we have seen that several aspects of sea ice kinematics (drift and deformation) cannot be explained by the properties of the atmospheric and oceanic forcings, and instead argue for a strongly non-linear, brittle behavior of the sea ice cover. In this chapter, we will review the scaling properties of sea ice fracturing and fragmentation, and see how these brittle processes can accommodate most of the sea ice deformation. Before this, however, an excursion into sea ice internal stresses will be considered. Internal stress states and their fluctuations through time are key features that characterize the mechanical behavior of sea ice (Weiss and Marsan 2004; Weiss et al. 2007). In addition, internal stresses enter the momentum equation of sea ice (Eq. 1.1) through their divergence $\nabla \cdot \sigma$ and so are expected to partly control ice dynamics (see also Sect. 2.5).

4.1 Data

In this chapter, two very different types of data are discussed: in situ internal stress measurements on the one hand, and fracture, fault and fragmentation patterns as seen in aerial or satellite images on the other hand.

Local, in situ ice stress measurements within the ice pack are available for more than 25 years (Cox and Johnson 1983) and several field campaigns have been described (Richter-Menge and Elder 1998; Richter-Menge et al. 2002; Tucker and Perovich 1992). Data were obtained from vibrating wire sensors allowing an estimation of the 2D (horizontal) stress tensor at a given point of the ice pack. The accuracy of instantaneous measurements is 20 kPa for stress amplitude and 5° for the principal stress directions (Cox and Johnson 1983). The longest records, sampled at a frequency of 1 h, have been obtained during the SIMI (Richter-Menge and Elder 1998) and SHEBA (Richter-Menge et al. 2002) field campaigns, and lasted about one Arctic winter (from October to April). Owing to the very large aspect ratio of lateral extension to thickness of the sea ice cover, the plane stress hypothesis is relevant and so the internal stress state is correctly estimated by such 2D sensors. An analysis of the vertical structure of this stress state can however be obtained from the superposition of a few sensors at the same location (e.g. Tucker and Perovich 1992).

Fracture and fragmentation patterns can be analyzed from aerial or satellite images in the visible spectrum. Owing to the very large albedo difference between snow-covered sea ice and open water/newly formed thin ice, fractures are easily identified, and associated patterns are retrieved from simple binarization algorithms on greyscale images (e.g. Weiss and Marsan 2004 and Fig. 4.1). The spatial resolution obviously depends on the altitude and/or the instrument: it can be well below 1 m for aerial images, a few meters to tens of meters for satellite imagery such as SPOT (Système Pour l'Observation de la Terre) or Landsat, and several hundred meters for MODIS (Moderate-Resolution Imaging Spectroradiometer). A combination of several data sources is therefore useful to analyze fracturing processes across scales (Schulson 2004). The main limitation of these instruments working in the visible spectrum is their uselessness during winter and/or in the absence of clear sky. An algorithm has been recently proposed to detect sea ice leads from passive microwave imagery, thus getting round this limitation, with partial success (detection of leads larger than 3 km, 50 % of total leads are detected) (Röhrs et al. 2012). As long as the ice/snow surface is dry, SAR images show a excellent contrast between sea ice and open water (fractures). This could be

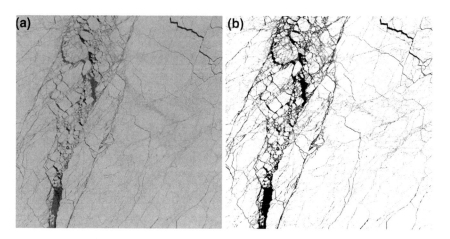

Fig. 4.1 **a** SPOT satellite greyscale image of the sea ice cover taken 6 April 1996 (i.e. early spring), centred around N 80°11′, W108°33′ and covering 60 × 60 km^2 with a resolution of 10 m/pixel. **b** Binarized version of the image, highlighting fractures and faults. A large (few km wide) fault with fragmented material inside is clearly visible on the *left* part of the image

used to obtain fracture patterns during winter, but the author is not aware of such analysis in the literature. Note also that strain-rate fields, such as those obtained from SAR data (Fig. 3.1), might be thresholded to extract patterns of active faults, at the cost of information lost compared to the scaling analysis of the entire field (Sect. 3.2).

4.2 Sea Ice Internal Stresses, Strength, and Rheology

A first inspection of in-situ stress records reveals several important characteristics (Richter-Menge and Elder 1998; Richter-Menge et al. 2002; Tucker and Perovich 1992).

 (i) The principal stresses (σ_I, σ_{II}), or the corresponding stress invariants (e.g. maximum shear stress $\tau = (\sigma_I - \sigma_{II})/2$) appear qualitatively highly intermittent, with periods of strong activity that can last from a few hours to several days, separated by quiescent periods (Fig. 4.2a).
 (ii) The maximum compressive stresses recorded during these episodes of activity are a few hundreds of kPa at most. Tensile stress states are present, but rarely exceed 50 kPa in magnitude.
(iii) Sea ice internal stresses can be decomposed into (a) thermal stresses associated with variations of air temperature (Lewis 1998) and (b) stresses related to external mechanical forcing and ice motion, such as wind forcing or collisions between adjacent floes (Lewis and Richter-Menge 1998). The thermal stress states are isotropic, either tensile or compressive, i.e. the "pressure" $\sigma_N = (\sigma_I + \sigma_{II})/2$ fluctuates but the shear stress $\tau = (\sigma_I - \sigma_{II})/2$ is negligible (Richter-Menge et al. 2002; Weiss et al. 2007). Thermal stresses can oscillate during spring with a daily periodicity (Hutchings et al. 2010). Note that a twice-daily periodicity, possibly tidal or inertially induced, has been also reported in some cases (Tucker and Perovich 1992). The consequence of this stress decomposition is that the shear stress invariant τ can be considered as a signature of ice motion-related stresses, and its analysis will be featured in what follows.
(iv) In general, stresses appear more intermittent and of larger amplitude near the edges of ice floes compared to the interiors. A significant part of the variability is truly local, as the correlation between shear stress records obtained from sensors placed a few hundred meters away on the same floe can vary from a very small correlation (correlation coefficient $R < 0.1$) to a rather significant one ($R > 0.6$) (Richter-Menge et al. 2002). However, similar correlations are also observed between sensors placed on different floes, meaning that another part of the variability has a more regional origin.

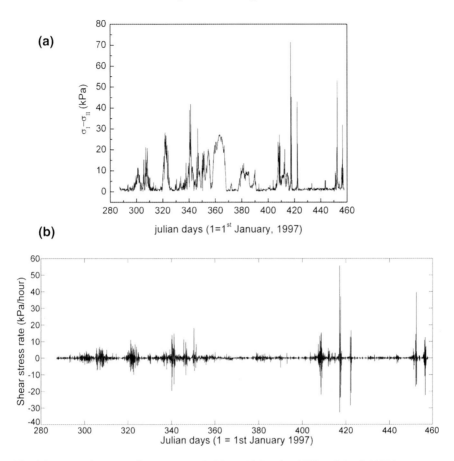

Fig. 4.2 a Ice "shear stress" $\sigma_I - \sigma_{II}$ recorded from 14 October 1997 to 2 April 1998 by a sensor during the SHEBA program. σ_I and σ_{II} are the 1st and 2nd principal stresses. The signal plotted is actually twice the shear stress $\tau = (\sigma_I - \sigma_{II})/2$. From Weiss and Marsan (2004). **b** Shear stress rate (time derivative of the signal of Fig. 4.2a), calculated at the sampling frequency of 1/h

(v) The few SHEBA data recorded during summer show very little internal stress during this season, in agreement with a less concentrated, more fragmented ice pack. Indeed, the SHEBA experiment was located near the boundary between the perennial and the seasonal ice cover, i.e. a region where mechanical interactions between floes are expected to almost vanish during summer. This might be different well inside the perennial ice pack.

Beyond these observations, more quantitative analyzes of these data are enlightening for sea ice mechanics and rheology. Represented in a Mohr-Coulomb diagram of the maximum shear stress τ *versus* pressure σ_N, the recorded stress states lie within an envelope with clear Coulombic boundaries described by:

$$|\tau| = |\tau_0| - \mu\sigma_N \qquad (4.1)$$

where the cohesion $|\tau_0|$ lies in the range 30–60 kPa and $\mu = 0.7 \pm 0.1$ is a "friction" coefficient (Weiss et al. 2007) (Fig. 4.3). However, as the angle between the plane of maximum shear stress and the plane of failure is unknown in this case, μ cannot be interpreted directly as an internal friction coefficient. Assuming that the fault orientation maximizes the so-called Coulomb stress $|\tau_0| - \mu\sigma_N$, an internal friction coefficient $\mu_i = 0.9$ can be deduced. Note also that this internal friction cannot be formally identified as a friction between adjacent surfaces, although the phenomenology of Eq. (4.1) is similar to that of static friction (See e.g. Jaeger and Cook 1979; Weiss and Schulson 2009 for more details about Coulombic rheology and internal friction). The implications of this are important.

Firstly, such Coulombic envelopes argue for a deformation of the sea ice cover essentially accommodated by faulting processes involving brittle failure and friction, in agreement with the premises discussed in Chap. 3. Secondly, the similarity with failure envelopes obtained from laboratory mechanical tests on ice samples (Schulson et al. 2006a, b) is striking (Weiss and Schulson 2009): lab-scale envelopes are also defined by Coulombic branches with an internal friction coefficient that slightly varies with the loading rate or temperature (Fortt and Schulson 2009; Schulson et al. 2006a), but lies in the range 0.7–1.0, i.e. is in excellent agreement with the "field" coefficient. This means that the underlying Coulombic faulting mechanism is scale-invariant (Weiss and Schulson 2009).

While the internal friction coefficient appears to be scale invariant, the failure strength is not. In the Mohr–Coulomb plane (Fig. 4.3), the strength is represented by the cohesion $|\tau_0|$, which can be understood as a shear strength in the absence of normal stress. Note that an alternative representation of the Coulombic envelope in the principal stress space is $\sigma_I = A\sigma_{II} + \sigma_c$, where σ_c can be interpreted as a

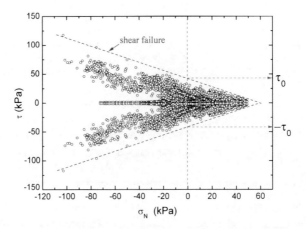

Fig. 4.3 Stress states recorded during the Arctic winter of the SHEBA program at the sensor "Baltimore" from mid-October, 1997 to the end of June, 1998 (1 measurement per hour), plotted in a τ versus σ_N graph. *The dotted red line* represents $|\tau| = |\tau_0| - \mu\sigma_N$, with $|\tau_0| = 40$ kPa and $\mu = 0.7$, corresponding to shear failure. *The dotted gray lines* show the cohesion value, $\pm\tau_0$. (Adapted from Weiss et al. 2007)

4.2 Sea Ice Internal Stresses, Strength, and Rheology

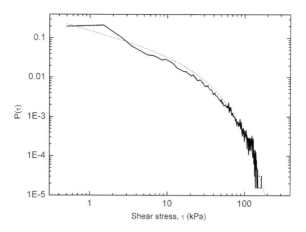

Fig. 4.4 Probability density distribution of the shear stress states, $\tau = (\sigma_I - \sigma_{II})/2$, recorded during the winter season (October to April) of the SHEBA program (*thick line*). This PDF is built from a compilation of the data of all the SHEBA stress sensors. Truncated power law fit, $P(|\tau|) \sim |\tau|^{-\eta} \exp\left(\frac{-|\tau|}{|\tau_c|}\right)$, with $\eta = 0.6$ and $|\tau_c| = 27$ kPa (*thin dashed line*)

compressive strength with $\sigma_c = \frac{2|\tau_0|}{1+\mu}$ and $A = \frac{1-\mu}{1+\mu}$. The in-situ sea ice strength (30–60 kPa for the cohesion, corresponding to 35–70 kPa for σ_c) appears therefore much lower than experimental values obtained from saline ice samples in the lab (~ 5 MPa for uniaxial compressive strength (Schulson and Duval 2009; Schulson et al. 2006b), i.e. ~ 4 MPa for cohesion). Before discussing the origin of the scale effect on strength, this raises the important question: Can we unambiguously define a failure strength for the sea ice cover?

To answer this question, a statistical analysis is required. Figure. 4.4 shows the PDF of a compilation of the shear stresses $|\tau|$ recorded at a frequency of 1 h by all the stress sensors installed during the SHEBA field campaign, from October 1997 to June 1998. The SIMI field data give similar results. This distribution is well fitted by a truncated power law, $P(|\tau|) \sim |\tau|^{-\eta} \exp\left(\frac{-|\tau|}{|\tau_c|}\right)$, where the exponential tail defines a characteristic value $|\tau_c| = 27$ kPa, slightly below the cohesion obtained in Fig. 4.3. This means that an upper bound exists for the shear stress PDF, and a value of ~ 30 kPa can indeed be considered a correct estimate of the failure strength of pack ice in winter. On the reverse, a pure power law without exponential tail would mean that stresses can reach arbitrarily large values. The scarce shear stress data recorded during summer at SHEBA indicates a nearly vanishing strength in a much looser ice cover.

Thus to explain the scale effect on strength, simple arguments can be considered (Schulson 2004; Weiss et al. 2007): in classical fracture mechanics, the failure strength depends on the size l of the defect, or "stress concentrator", whose activation will eventually lead to failure, as $l^{-1/2}$. Therefore, for a scale effect on shear strength of $\frac{1\,\mathrm{MPa}}{30\,\mathrm{kPa}}$, one gets $\frac{l_{field}}{l_{lab}} = \left(\frac{1000}{30}\right)^2 \approx 10^3$. For a size of stress concentrator of the order of the grain-size (mm to cm) for laboratory tests on undamaged samples, the above scaling relation leads to l_{field} of the order of magnitude of the ice cover's thickness. Although necessarily rough, this estimation suggests the ubiquity of planes of weakness at the scale of the ice cover's thickness, possibly related to structural heterogeneities or to small thermal cracks. If a larger pre-existing fault is activated and/or the stress sensor is far from it, the recorded stress magnitude should be smaller, while the shape of the failure envelope (Fig. 4.3) should be preserved. Such reduced but homothetic envelopes have been identified in internal stress records (Weiss et al. 2007).

These analyzes lead Weiss et al. (2007) to propose a qualitative scenario where most sea ice deformation is accommodated by various fracturing/faulting episodes over a broad range of spatial scales. Such a scenario was then refined and quantified by Marsan and Weiss (2010) and Weiss et al. (2009) (see Sect. 4.5).

4.3 Intermittency of Sea Ice Stresses

We now turn our attention to the most prominent characteristic of internal stress records, their intermittency (Fig. 4.2), which was explored by Weiss (2008), Weiss and Marsan (2004) and more recently by Hutchings et al. (2010). A spectral analysis reveals a power law scaling for the power spectrum of the shear stress τ over the timescales 1 h-100 days, $p_\tau(\omega) \sim \omega^{-\zeta}$, with ζ in the range 1.4–1.7 (Hutchings et al. 2010; Weiss and Marsan 2004) (Fig. 4.5). Principal stress records show a similar behavior, sometimes augmented by a peak at 1 day, a likely signature of thermal stresses induced by variations of air temperature with a daily periodicity (Hutchings et al. 2010), which is absent by construction in the shear stress invariant. As the power spectrum is the Fourier transform of the autocorrelation function of the signal, this scaling indicates a long-term memory of the stress record, confirmed by an autocorrelation analysis showing zero crossing around 1 month (not shown). On the other hand, such ζ values correspond to "roughness" exponents of the stress record $H = (\zeta-1)/2$ clearly below 0.5. This means that internal shear stress records are anti-persistent (see also Sect. 3.5), with

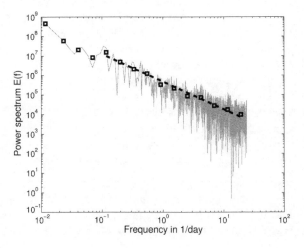

Fig. 4.5 Power spectrum of the shear stress record of Fig. 4.2a (in *light gray*). A power-law fit (*thick dashed line*) is estimated by averaging the spectrum over frequency windows with algebraically increasing width (*squares*), yielding $p(\omega) \sim \omega^{-\zeta}$, with $\zeta = 1.42$ (From Weiss and Marsan 2004)

shear stress increases followed generally by stress drops. This is actually the signature of a "vibrating" character of stress fluctuations, which becomes obvious when plotting the stress rates $d\tau/dt$ (Fig. 4.2b). These stress rate records have short memory when considering signed values, but a memory of a few days (~ 100 h) when considering absolute values. The resemblance of (Fig. 4.2b) with icequake records (Marsan et al. 2011) is questionable, especially because the characteristic period of these "bursts" seems very large (a few hours). This deserves further attention in the future.

A more complete multifractal characterization of the intermittency of ice motion-related stresses (shear) was performed by Weiss and Marsan (2004). They analyzed the temporal scaling of the stress increments $\langle|\Delta\tau|^q\rangle = \langle|\tau(t+\Delta t) - \tau(t)|^q\rangle$, and found a multifractal behavior $\langle|\Delta\tau|^q\rangle \sim \Delta t^{\varsigma(q)}$, with a non-linear moment function $\varsigma(q)$. The quantity $\langle|\Delta\tau|\rangle/\Delta t$ can be seen as a stress rate measured at the time scale Δt. It was found to scale as $\Delta t^{-0.66}$, as $\varsigma(1) = 0.34$, hence to increase with decreasing time scale. The moment function of the stress rate, $\varsigma(q) - q$, fully characterizes the intermittency of stress fluctuations such as those of Fig. 4.2b. As the higher moments of the stress rate distribution grow faster towards small scales than for a monofractal scaling, the stress record is increasingly "localized" in time towards small scales.

This intermittency is not restricted to stress or stress-rate amplitudes, but also characterizes the changes in principal stress directions (Weiss 2008). Figure. 4.6 shows a reconstructed (see Weiss 2008 for more details) principal stress direction time series for a SHEBA stress sensor from October 1997 to June 1998. This signal is punctuated by large rotations taking place over several hours. The rotational fluctuations $\Delta\theta$ are indeed distributed following strongly non-Gaussian PDFs with heavy tails. A spectral analysis demonstrated a power law scaling for the power spectrum of the stress direction records, $p_\theta(\omega) \sim \omega^{-\zeta}$, with $\zeta = 1.76$, significantly below the Brownian noise exponent $\zeta = 2$. This analysis also revealed a multifractal scaling of rotations, $\langle|\Delta\theta|^q\rangle = \langle|\theta(t+\Delta t) - \theta(t)|^q\rangle \sim \Delta t^{\varsigma(q)}$. These characteristics strongly differentiate sea ice principal stress direction statistics from wind directions, which are Brownian-like in the frequency domain with exponential PDFs (van Doorn et al. 2000; Weiss 2008). This shows clearly that the intermittency of stress fluctuations is not a direct inheritance of wind forcing. Instead, it likely results from the fracturing process itself: if a fractured elastic plate is loaded by a forcing whose principal directions change through time in a relatively random, Brownian-like manner, it will not respond simply and immediately to these random fluctuations of the forcing field because of the presence of pre-existing, persistent fractures that act as preferential paths to accommodate the deformation and control the elastic stress field. This is consistent with the persistence of active fractures with nearly constant orientations over long time scales (>1 month) (Coon et al. 2007), as well as with the sudden shifts of orientation that can be seen on the animations of satellite-derived strain-rate fields (Kwok 2001).

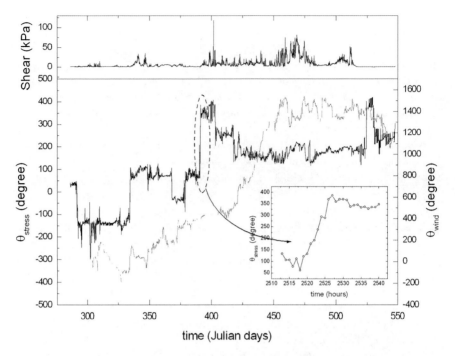

Fig. 4.6 *Top panel*: Evolution of the shear stress amplitude at the SHEBA Baltimore sensor, from 13 October 1997 to 1 June 1998. *Bottom panel*: Reconstructed principal stress (*black line*) and 2 m-wind (*gray line*) direction time series over the same period. The wind data were obtained at the Atmospheric Surface Flux Group (ASFG) tower (Persson et al. 2002). See Weiss (2008) for details about the reconstruction procedure. *Inset*: Zoom of the principal stress direction time series over a period of about 1 day showing a rotation of about 300° taking place in about 10 h (From Weiss 2008)

4.4 Fracture Networks

The analysis of internal stresses demonstrated the elasto-brittle rheology of the sea ice cover, and argued for a sea ice deformation accommodated by fractures and faults over a wide range of spatial and temporal scales (Weiss et al. 2007). Before building a statistical model relating brittle deformation to fracturing (Sect. 4.5), a preliminary question arises: Are sea ice fracture patterns indeed characterized by multiscale properties?

Historically, the first evidence of scaling for fracture and fragmentation of sea ice came from the analysis of floe size distributions. Rothrock and Thorndike (1984) gave clear qualitative examples of the scale invariance of fragmented sea ice, and, using aerial photographs and satellite images of the Arctic ocean with different resolutions, reported power law distributions of floe sizes s (defined as a mean floe diameter on 2D images $P(s) \sim s^{-\eta}$ with $\eta \approx 2.7$–2.8, over the scale range 0.1–50 km. Then, other authors confirmed this observation, with exponents

η in the range 2.0–3.0 and spatial scales in the range 0.1–50 km (Kergomard 1989; Lensu 1990; Matsushita 1985; Toyota et al. 2006; Weiss and Marsan 2004). For a compilation, see Weiss (2003). Such power law distributions have been extensively analyzed in the context of rock fragmentation and the formation of fault gouges (e.g. Keulen et al. 2007; Korvin 1992; Sammis and Biegel 1989; Sammis et al. 1987; Turcotte 1992), and discussed in terms of resulting friction (Biegel et al. 1989), or of the energy required to fragment the gouge (Chester et al. 2005; Keulen et al. 2007). Simple hierarchical models have been proposed to explain these observations, all based on a scale invariant fracturing process (Steacy and Sammis 1991; Turcotte 1986). The point to remember here is that such power law distribution of floe sizes is the fingerprint of an underlying scale-independent fracturing and fragmentation process of pack ice, from the scale of the ice cover's thickness ($\sim$ m) to the regional scale ($\sim$ 50–100 km) and possibly above.

This scale invariance, however, seems to break in the marginal ice zone (MIZ) defining the transition from pack ice to the open ocean. Lu et al. (2008) reported cumulative distributions (CDF) of floe sizes following truncated power laws of the form $P(>s) \sim s^{-\eta+1} - s_c^{-\eta+1}$, with a cut-off size s_c decreasing from the ice pack to the ice edge where it falls to $\sim$ 20 m. This breaking of scaling might be explained by a specific fracturing mechanism, namely the bending failure of the ice cover induced by ocean swell. This mechanism indeed introduces a maximum floe size, directly related to the swell's wavelength, and predicts an increase of the maximum floe size from the ice edge inward (Dumont et al. 2011), as observed (Lu et al. 2008). Far from the MIZ, the influence of this bending failure mode disappears, and scale invariance prevails. For situations corresponding to well separated floes, i.e. to ice concentrations below 50 % (open drift ice), clustering effects arising from loss of kinetic energy during collisions between floes, or re-aggregation by thin ice growth, may modify the floe size PDF (Herman 2011).

In a compact sea ice cover, floes are delimited by fractures and faults. Therefore, the power law distribution of floe sizes is a first expression of the scale invariance of sea ice fracturing mechanisms and networks. This scale invariance can be highlighted in different ways. Schulson (2004) found strong similarities between fracture features observed on biaxially compressed ice samples in the lab (mm to cm scales) and on aerial and satellite images (10^2–10^5 m scale), such as narrow lineaments with evidence of relative movement between crack faces, like strike-slip faults within the Earth's crust. Conjugate sets of such shear faults are observed both at the lab and geophysical scales, with an intersection angle 2φ that is related to the internal friction coefficient μ_i through $\tan(2\varphi) = \frac{1}{\mu_i}$. Lab values ($2\varphi \approx 50°$ (Schulson 2004), giving $\mu_i \approx 0.8$, are in excellent agreement with the internal friction coefficient deduced from Coulombic stress envelopes (Sect. 4.2), whereas estimated geophysical values are more widespread [30°–50° (Erlingsson 1988; Schulson 2004)]. Considering scale invariant fracturing/faulting mechanisms, the rheology of sea ice can be discussed from scaling arguments and the application of mechanical models initially developed at the laboratory scale. In Sect. 3.4, we compared estimated sea ice strain-rates directly with the ductile-to-brittle transition strain rate $\dot{\varepsilon}_{D/B}$ obtained from laboratory experiments ($\approx 10^{-4}$ s^{-1} ≈ 10 day^{-1}), without considering any specific mechanism or the presence of defects within the material. Schulson (1990, 2001) examined this transition under multiaxial compression and proposed that it occurs when the rate of stress build-up at defects of length l (playing the role of "stress concentrators") exceeds the rate of stress relaxation through creep. In this formulation, the transition strain-rate $\dot{\varepsilon}_{D/B}$ scales as $l^{-1.5}$. This means that the larger the stress concentrator, the lower the transition strain-rate. Considering lab values of $\dot{\varepsilon}_{D/B} = 10$ day^{-1} and $l \approx 1$ cm (grain-size), as well as a "field" concentrator of ~ 1 m (the ice cover's thickness), one then deduces from this scaling a transition strain-rate for sea ice around 10^{-2} day^{-1}, i.e. a value exceeded in most RGPS 10 km cells (see Chap. 3 and Figs. 3.2 and 3.3). Larger stress concentrators would be associated with even lower transition strain-rates (Schulson 2004). These simple calculations reinforce the hypothesis of a sea ice deformation accommodated by fracturing/faulting events over a broad range of spatial scales, via the operation of scale-independent mechanisms (Schulson 2004).

A more quantitative analysis of the scale invariance of sea ice fracture networks was performed by Weiss and Marsan (2004) from a SPOT image (Fig. 4.1). They found (i) that the fracture network defined a power law distribution of floe sizes, and (ii) that the fraction of open water (the "fracture density") p_L measured at scale L followed a multifractal scaling

$$\langle p_L^q \rangle \sim L^{-\chi(q)} \tag{4.2}$$

over at least 2.5 orders of magnitude (0.2 km $\leq L \leq$ 60 km = SPOT image size), with $\chi(q)$ being a non-linear, convex function of order q. Note that in this case of a conservative scalar variable, the value of $\chi(1)$ is trivial, $\chi(1) = 0$. The similarity between the multifractal scaling of the fracture density (Eq. 4.2), and of the sea ice strain-rates (Eq. 3.3), is obviously not coincidental, but instead argues further for the brittle nature of sea ice deformation.

To conclude this section, it is worth considering another signature of scale invariance of fracture networks, more specific than the scaling of fracture density just discussed, but of surprisingly strong importance in terms of ocean–atmosphere interactions. As mentioned in the beginning of this chapter, leads (fractures) concentrate the latent and sensible heat fluxes between the ocean and the atmosphere. However, narrow leads (several m) are much more efficient at transmitting turbulent heat than larger ones (several hundreds of m and above) (Alam and Curry 1997; Andreas and Cash 1999). Lead widths w are power-law distributed, $P(w) \sim w^{-\eta}$, with η in the range 1.6–2.6, down to very small spatial scales (<20 m) (Lindsay and Rothrock 1995; Marcq and Weiss 2012). This implies that the open water fraction is by far dominated by very narrow, thus efficient leads for the heat exchange point of view. Therefore, for a given open water fraction of a few percent, neglecting such a lead width distribution leads to a strong underestimation of heat fluxes (Marcq and Weiss 2012). This example illustrates the importance of sea ice brittle deformation and associated scaling laws for the evolution of the ice cover and of the Arctic climate. Other evidence will be presented in the conclusion of this volume (Sect. 5.3).

4.5 A Statistical Model of Sea Ice Fracturing and Deformation

The observations and analyzes reviewed above argue for sea ice deformation resulting from the combination of various short and transient fracturing/faulting episodes occurring along narrow cracks/faults over a wide range of space and time scales, in a way similar to the brittle deformation of the upper crust. If so, can we build a multiscale statistical model of brittle deformation based on the superposition of discrete displacements that would allow us to reproduce the scaling properties detailed in Chap. 3, including the space/time coupling?

Once again, Thorndike (1986a) was the first to propose such a statistical model where sea ice deformation results from the collective effect of many leads/fractures. This was an important conceptual step, as it considered sea ice deformation to be discontinuous by nature. This model was however based on crude hypotheses, such as a fully random (Poisson) distribution of fracturing events in space and time, and a Gaussian distribution for the displacements across the leads. These assumptions are not compatible with our present knowledge of sea ice deformation and fracturing, such as the scaling properties of sea ice deformation in space and time (Sects. 3.2 and 3.3), or the power law distribution of strain-rates (Sect. 3.2) as well as of lead openings (Sect. 4.4). Consequently, Weiss et al. (2009) and Marsan and Weiss (2010) elaborated on this seminal model, taking into account these scaling properties, and building analogies with the crustal deformation accommodated by earthquakes.

This model considers that nearly all the sea ice deformation is accommodated by multiple fracturing and faulting events at various scales, i.e. the contribution of dislocation-related creep is negligible (Weiss et al. 2007), and the elastic part of the deformation is neglected, an approximation that is reasonable when considering length scales greater than the thickness of the ice cover. To each event of length l is associated an offset w, averaged along l. In this simple scalar model, the fracture modes are undifferentiated, i.e. w can represent an opening for a tensile fracture, a shear offset for a shear fault, or convergence during a ridging event. Here, w represents the offset resulting from a brief, transient episode of fracturing, and not the total displacement that could accumulate along a fault over a long time as the result of multiple episodes. In agreement with the aspect ratio of the ice cover, the model is 2D, i.e. the thickness of the fractures is considered to be constant and corresponds to the ice cover thickness. The model is isotropic and does not consider fracture orientations. With these hypotheses, the strain increment ε induced by a fracturing event over a region of size L (i.e. of area L^2) is:

$$\varepsilon = \frac{w}{l} \times \frac{l^2}{L^2} = \frac{wl}{L^2} \qquad (4.3)$$

Following earlier work on faulting of the Earth's crust (e.g. Scholz and Cowie 1990), the offset w is assumed to be proportional to l, $w \sim l$, with a proportionality constant around 10^{-5} for the Earth's crust (Wells and Coppersmith 1994). This proportionality is compatible with fracture mechanics principles, but the constant is unknown for the sea ice cover and possibly depends on the driving stress.

The model then builds on three additional hypotheses inspired from earthquake phenomenology and compatible with sea ice observations:

(i) Instead of considering a random spatial distribution of events as Thorndike (1986), a fractal fracture pattern is assumed, characterized by a number of event barycenters within a region of area L^2, $N(L)$, scaling as.

$$N(L) \sim L^{D(t)} \qquad (4.4)$$

where $0 \leq D(t) \leq 2$ is a fractal dimension that increases while increasing the time scale considered. This relation is actually the mirror expression of Eq. (3.6), formulated in terms of discrete brittle deformation events (Marsan and Weiss 2010). Such spatial scaling has been documented for earthquakes (Kagan 1991; Kagan and Knopoff 1980). The dependence of D on time scale t expresses a memory effect of the system that slowly weakens with time. At short time scales, fracturing/faulting events are strongly clustered ($D \ll 2$) as the result of elastic interactions and stress redistributions. On the other hand, as $D(t \to +\infty) \to 2$, being on a fracture at time $t = 0$ reveals nothing about the initiation of fractures nearby at $t = \infty$ (years). In other words, this illustrates a space/time coupling of brittle deformation, as discussed in Sect. 3.4. Such a fractal pattern of fracturing events is also in qualitative agreement with the fractal character of fracture networks discussed in the preceding section.

(ii) Similarly, for a given region of area L^2, the deformation over this region during the time t results from the cumulative effect of $N(t)$ independent events (i.e. the size of event of rank n is independent of the size of events $n - 1$ or $n + 1$), with $N(t)$ following:

$$N(t) \sim t^{\delta(L)} \qquad (4.5)$$

where $0 \leq \delta(L) \leq 1$ is an exponent depending on the spatial scale considered. As Eq. 4.4 was the mirror expression of Eq. (3.6), this relation is the mirror of Eq. (3.7), i.e. it expresses intermittency of the deformation process in terms of discrete events. The boundary value $\delta = 0$ would correspond to a fracturing history boiling down to an isolated event, a situation only recovered when considering $L \to 0$, whereas $\delta = 1$ represents the Poisson hypothesis of purely random, uncorrelated events, i.e. the situation assumed by Thorndike (1986a). This kind of scaling has been reported for earthquakes (Kagan and Jackson 1991). It is in this case an alternative way to formulate the intermittency or the time clustering of events caused by earthquake interaction mechanisms, i.e. the triggering of aftershocks by previous earthquakes generally expressed by Omori's law (Omori 1894). It is also in agreement with the temporal correlation between icequakes recorded at seismic stations positioned around the sailboat *Tara* during its journey along the transpolar drift in 2007 (Marsan et al. 2011) (Fig. 4.7), although in this case the positioning of icequakes was not possible from the seismic network.

(iii) Instead of Thorndike's (1986) hypothesis of a Gaussian distribution of the displacements across the fractures, i.e. mild randomness, Weiss et al. (2009) and Marsan and Weiss (2010) considered a power law distribution of offsets, $P(w) \sim w^{-\eta}$, i.e. wild randomness, in agreement with the power law distribution of lead widths (Sect. 4.4). As shown by Marsan and Weiss (2010), this is also compatible with a power law distribution of seismic moments M, as observed for earthquakes. M is proportional to the offset w times the rupture area, i.e. scales as $M \sim whl \sim w^2$ for the sea ice cover. For earthquakes, $P(M) \sim M^{-(1+\frac{2b}{3})}$, where the b-value characterizes the cumulative distribution of earthquake magnitudes [the Gutenberg-Richter distribution (Gutenberg and Richter 1954)]. This would correspond to $\eta = 1 + \frac{4b}{3}$ for the distribution of offsets w (Marsan and Weiss 2010). For crustal seismicity, b is close to 1 (Utsu 2002), which would lead to $\eta = \frac{7}{3}$, a value in full agreement with recently obtained lead width distributions (Marcq and Weiss 2012). Such power law distribution of seismic moments is also compatible with a power law distribution of recorded icequake "intensities" (Marsan et al. 2011 and Fig. 4.7), although the seismic moments were not defined in this case as the quake hypocenters were not determined.

The combination of such power law distribution of offsets with Eqs. 4.3–4.5 allows us to map this statistical model of discrete fracturing events onto a continuum deformation picture, and to recover the scaling laws of sea ice deformation (Weiss et al. 2009). More specifically, taking $\eta = \frac{7}{3}$ (i.e. $b = 1$), the following correspondence can be found in the spatial domain (Weiss et al. 2009)

$$\beta(t) = 2 - \frac{3}{2}D(t) \qquad (4.6)$$

At the time scale of a few hours, $\beta = 0.85$ for both winter and summer (Sect. 3.3 and Rampal et al. 2008). This corresponds to $D \approx 0.77$, indicating that at these short time scales the fracturing events are strongly clustered. This is however significantly above the boundary value $D = 0$ corresponding to an isolated event, suggesting that the transient fracturing episodes occur at much shorter time scales than a few hours. At long time scales (months), $\beta = 0.35$ in winter and 0.42 in summer, corresponding respectively to $D = 1.1$ and 1.05. This is significantly below the boundary value $D = 2$ corresponding to a dense, homogeneously distributed (Poisson) network, and suggests that fracturing events are still organized along 1D-like narrow structures at these time scales. This is another way to express the heterogeneous character of sea ice deformation and fracturing up to very large time scales.

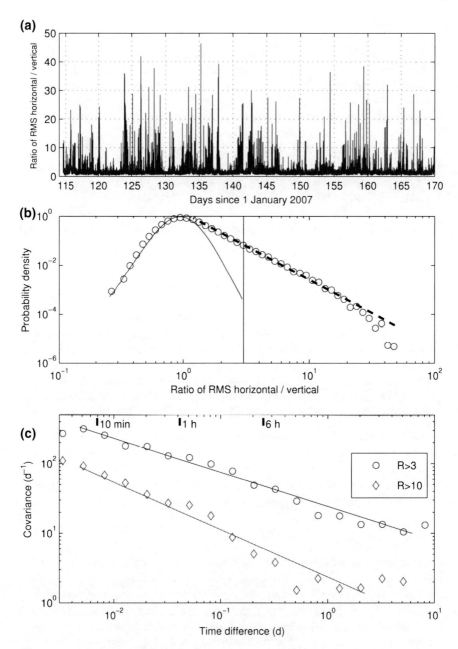

Fig. 4.7 **a** Low frequency icequakes recorded by a seismic station positioned near the sailboat *Tara* during the DAMOCLES experiment in spring 2007. These icequakes are characterized by a low characteristic frequency (typically 0.025 Hz) and by horizontal motion only. Consequently, their "intensity" can be calculated by the ratio R between the RMS of a horizontal channel and the vertical channel of the station. **b** Probability density function (PDF) of R. For $R > 1$, the PDF decays as a power law, $R^{-2.7}$ (*dashed line*). **c** Temporal correlation for $R > 3$ and $R > 10$ events, showing a power law decay of $\Delta t^{-0.48}$ and $\Delta t^{-0.68}$, respectively. (From Marsan et al. 2011)

4.5 A Statistical Model of Sea Ice Fracturing and Deformation

In the time domain (Weiss et al. 2009),

$$\alpha(L) = 1 - \frac{3}{2}\delta(L) \qquad (4.7)$$

At small spatial scales (1 km), $\alpha = 0.89$ for winter (0.87 for summer, see Sect. 3.3 and Rampal et al. (2008)), which leads to $\delta \approx 0.07$. This value is small, meaning that at these small spatial scales, the deformation process is highly intermittent with bursts of events separated generally by long quiescent episodes, in agreement with seismic data (Marsan et al. 2011). At very large spatial scales (a few hundred km), $\alpha = 0.30$ for winter (0.25 for summer) corresponding to $\delta = 0.47$ (0.5 for summer). This is significantly below the boundary value $\delta = 1$, meaning that some intermittency remains even at the regional scale.

The mirror expression for the space/time coupling of sea ice deformation Eq. (3.8), in terms of discrete events, becomes (Marsan and Weiss 2010)

$$N(L, t) \sim t^{\delta_0} L^{D_0} e^{c' \ln(t) \ln(L)} \qquad (4.8)$$

with $c' = \frac{8}{3}c$. It has been shown that such space/time coupling symmetry is also relevant for crustal seismicity (Marsan and Weiss 2010).

In conclusion, a simple statistical model elaborated on the seminal idea of a discrete, i.e. brittle nature of sea ice deformation proposed by Thorndike (1986a), but taking into account an intermittency and spatial clustering of fracturing/faulting events as well as the power law scaling of lead offsets, is able to reproduce the scaling properties of sea ice deformation in space and time, described in Chap. 3 from a continuous approach. This reinforces the view of the intrinsically brittle character of sea ice deformation already formulated and argued in the preceding sections of this chapter. It is also interesting to note that such a simple model can be used to estimate to what extent small *versus* large fractures/events contribute to the global deformation (Weiss et al. 2009). Considering Eq. 4.3 as well as the power law scaling of offsets w [taking $\eta = \frac{7}{3}$, i.e. $b = 1$ as above, whereas a value of $b = 1.25$ was originally considered in Weiss et al. (2009)], it can be shown that fracturing events longer than about $L_{max}/3$ account for half of the total deformation, where $L_{max} \approx 2000$ km is the size of the considered domain, i.e. the Arctic basin. This strikingly illustrates the role of the very large deformation events, spanning a significant part of the Arctic and easily visible on RGPS-based deformation maps (see e.g. Fig. 3.1).

References

Alam, A., & Curry, J. A. (1997). Determination of surface turbulent fluxes over leads in Arctic sea ice. *Journal of Geophysical Research, 102*(C2), 3331–3343.

Andreas, E. L., & Cash, B. A. (1999). Convective heat transfer over wintertime leads and polynyas. *Journal of Geophysical Research, 104*(C11), 25721–25734.

Andreas, E. L., & Murphy, B. (1986). Bulk transfer-coefficients for heat and momentum over leads and polynyas. *Journal of Physical Oceanography, 16*(11), 1875–1883.

Biegel, R. L., Sammis, C. G., & Dieterich, J. H. (1989). The frictional-properties of a simulated gouge having a fractal particle distribution. *Journal of Structural Geology, 11*(7), 827–846.

Chester, J. S., Chester, F. M., & Kronenberg, A. K. (2005). Fracture surface energy of the Punchbowl fault, San Andreas system. *Nature, 437*(7055), 133–136.

Comiso, J. C., Parkinson, C. L., Gersten, R., & Stock, L. (2008). Accelerated decline in the Arctic sea ice cover. *Geophysical Research Letters, 35*, L01703.

Coon, M., Kwok, R., Levy,G., Pruis, M., Schreyer, H. & Sulsky, D. (2007). Arctic ice dynamics joint experiment (AIDJEX) assumptions revisited and found inadequate. *Journal of Geophysical Research, 112*, C11S90.

Cox, G. F. N., & Johnson, J. B. (1983). Stress measurements in ice, *Rep. (83–23)*, CRREL, Hanover.

Dumont, D., Kohout, A., & Bertino, L. (2011) A wave-based model for the marginal ice zone including a floe breaking parameterization. *Journal of Geophysical.Research-Oceans, 116*.

Erlingsson, B. (1988). Two-dimensional deformation patterns in sea ice. *Journal of Glaciology, 34*, 301–308.

Fichelet, T., & Morales Maqueda, M. A. (1995). On modelling the sea-ice-ocean system. *Modélisations couplées en climatologie, 76*, 343–420.

Fortt, A. L., & Schulson, E. M. (2009). Velocity-dependent friction on Coulombic shear faults in ice. *Acta Materialia, 57*(15), 4382–4390.

Gutenberg, B., & Richter, C. F. (1954). *Seismicity of the Earth and associated phenomenon*. Princeton: Princeton University Press.

Heil, P., & Hibler, W. D. I. (2002). Modeling the high frequency component of arctic sea ice drift and deformation, *Journal of Physical Oceanography, 32*, 3039–3057.

Herman, A. (2011). Molecular-dynamics simulation of clustering processes in sea-ice floes. *Physical Review E, 84*(5).

Hutchings, J. K., Geiger, C. A., Roberts, A., Richter-Menge, J. A., & Elder, B. (2010). *On the Spatial and Temporal Characterization of Motion Induced Sea Ice Internal Stress*, paper presented at ICETECH10 (pp. 20–23). Alaska: Anchorage.

Jaeger, J. C., & Cook, N. G. W. (1979). *Fundamentals of Rock Mechanics*. London: Chapman and Hall.

Kagan, Y. Y. (1991). Fractal dimension of brittle fracture. *Journal Nonlinear Science, 1*, 1–16.

Kagan, Y. Y., & Jackson, D. D. (1991). Long-term earthquake clustering. *Geophysical Journal International, 104*, 117–133.

Kagan, Y. Y., & Knopoff, L. (1980). Spatial distribution of earthquakes: the two point correlation function. *Geophysical Journal Research Astron Socity, 62*, 303–320.

Kergomard, C. (1989). Analyse morphométrique de la zone marginale de la banquise polaire au nord-ouest du spitsberg à partir de l'imagerie SPOT panchromatique, *Bull. Secure File Transfer Protocol, 115*, 17–20.

Keulen, N., Heilbronner, R., Stuenitz, H., Boullier, A. M., & Ito, H. (2007). Grain size distributions of fault rocks: A comparison between experimentally and naturally deformed granitoids. *Journal of Structural Geology, 29*(8), 1282–1300.

Korvin, G. (1992). *Fractal models in the Earth Sciences*. Amsterdam: Elsevier.

Kwok, R. (2001). *Deformation of the arctic ocean sea ice cover between November 1996 and april 1997: A survey*, paper presented at IUTAM scaling laws in Iie mechanics and Iie dynamics. Fairbanks: Kluwer Academic Publishers.

Lemke, P., and others (2007). Observations: Changes in snow, ice and frozen ground, in *Climate Change 2007:* In S. Solomon, D. Qin, M. Manning, Z. Chen, M. Marquis, M. Averyt, M. Tignor and H. L. Miller. (Eds.)*The physical basis. Contribution of working group I to the fourth assessment report of the Intergovernmental Panel on Climate Change*.Cambridge: Cambridge University Press.

Lensu, M. (1990). *The fractality of sea ice cover*. Paper presented at IAHR Ice Symposium. Finland: Espoo.

References

Lewis, J. K. (1998). Thermomechanics of pack ice. *Journal of Geophysical Reserach, 103*(C10), 21869–21882.

Lewis, J. K., & Richter-Menge, J. A. (1998). Motion-induced stresses in pack ice. *Journal of Geophysical Research, 103*(C10), 21831–21843.

Lindsay, R. W., & Rothrock, D. A. (1995). Arctic sea ice leads from advanced very high resolution radiometer images. *Journal of Geophysical Research, 100*(C3), 4533–4544.

Lu, P., Li, Z. J., Zhang, Z. H., & Dong, X. L. (2008). Aerial observations of floe size distribution in the marginal ice zone of summer Prydz Bay. *Journal of Geophysical Research-Oceans, 113*(C2).

Marcq, S., & Weiss, J. (2012). Influence of sea ice lead-width distribution on turbulent heat transfer between the ocean and the atmosphere. *The Cryosphere, 6*(1), 143–156.

Marsan, D., & Weiss, J. (2010). Space/time coupling in brittle deformation at geophysical scales. *Earth and Planetary Science Letters, 296*(3–4), 353–359.

Marsan, D., Weiss, J., Metaxian, J. P., Grangeon, J., Roux, P. F., & Haapala, J. (2011). Low-frequency bursts of horizontally polarized waves in the Arctic sea-ice cover. *Journal of Glaciology, 57*(202), 231–237.

Matsushita, M. (1985). Fractal viewpoint of fracture and accretion. *Journal Physical Society Japan, 54*(3), 857–860.

Maykut, G. A. (1982). Large scale heat exchange and ice production in the central Arctic *Journal Geophysical Research, 87*(C10), 7971–7984.

Maykut, G. A. (1986). The surface heat and mass balance. In N. Untersteiner (Ed.), *The geophysics of sea ice* (pp. 395–464). New York: Plenum Press.

Nye, J. F. (1973). Is there any physical basis for assuming linear viscous behavior for sea ice. *AIDJEX Bulletin, 21*, 18–19.

Nye, J. F. (1975). The use of ERTS photographs to measure the movement and deformation of sea ice. *Journal of Glaciology, 15*(73), 429–436.

Omori, F. (1894). On the aftershocks of earthquakes. *Journal of the College of Science Imperial University of Tokyo, 7*, 111–120.

Persson, P. O. G., Fairrall, C. W., Andreas, E. L., Guest, P. S., & Perovich, D. K. (2002). Measurements near the atmospheric surface flux group tower at SHEBA: Near-surface conditions and surface energy budget. *Journal of Geophysical Research, 107*(C10), C000705.

Rampal, P., Weiss, J., Marsan, D., Lindsay, R., & Stern, H. (2008). Scaling properties of sea ice deformation from buoy dispersion analyses. *Journal of Geophysical Research, 113*, C03002.

Richter-Menge, J. A., & Elder, B. C.(1998). Characteristics of pack ice stress in the Alaskan Beaufort Sea. *Journal of Geophysical Research, 103*(C10), 21817–21829.

Richter-Menge, J. A., McNutt, S. L., Overland, J. E., & Kwok, R. (2002). Relating arctic pack ice stress and deformation under winter conditions. *Journal of Geophysical Research, 107*(C10), C000477.

Röhrs, J., Kaleschke, L., Bröhan, D., & Siligam, P. K. (2012). An algorithm to detect sea ice leads using AMSR-E passive microwave imagery. *The Cryosphere, 6*, 343–352.

Rothrock, D. A., & Thorndike, A. S. (1984) Measuring the sea ice floe size distribution. *Journal of Geophysical Research, 89*(C4), 6477–6486.

Sammis, C. G., & Biegel, R. (1989). Fractals, fault gouge, and friction, *PAGEOPH, 131*, 255–271.

Sammis, C. G., King, G., & Biegel, R. (1987). The kinematics of gouge deformation, *PAGEOPH, 125*(5), 777–812.

Scholz, C. H., & Cowie, P. A. (1990). Determination of total strain from faulting using slip measurements. *Nature, 346*, 837–839.

Schulson, E. M. (1990). The brittle compressive fracture of ice. *Acta Metallurgica et Materiali, 38*(10), 1963–1976.

Schulson, E. M. (2001). Brittle failure of ice. *Engineering Fracture Mechanics, 68*(17–18), 1839–1887.

Schulson, E. M. (2004). Compressive shear faults within the arctic sea ice: Fracture on scales large and small. *Journal of Geophysical Research, 109*, C07016, C002108.

Schulson, E. M., & Duval, P. (2009). *Creep and fracture of ice*. Cambridge: Cambridge University Press.

Schulson, E. M., Fortt, A. L., Iliescu, D., & Renshaw, C. E. (2006a). On the role of frictional sliding in the compressive fracture of ice and granite: Terminal vs post-terminal failure. *Acta Materialia, 54*, 3923–3932.

Schulson, E. M., Fortt, A. L., Iliescu, D., & Renshaw, C. E. (2006b) Failure envelope of first-year arctic sea ice: the role of friction in compressive failure. *Journal of Geophysical Research, 111*, C11S25.

Schulson, E. M., & Hibler, W. D. (1991). The fracture of ice on scales large and small: Arctic leads and wing cracks. *Journal of Glaciology, 37*, 319–322.

Schulson, E. M., & Hibler, W. D. I. (2004). Fracture of the winter sea ice cover on the Arctic ocean, *C.R. Physique, 5*, 753–767.

Serreze, M. C., & Francis, J. A. (2006). The arctic amplification debate. *Climate Change, 76*(3–4), 241–264.

Steacy, S. J., & Sammis, C. G. (1991). An automaton for fractal patterns of fragmentation. *Nature, 353*, 250–252.

Thorndike, A. S. (1986a). Diffusion of sea ice. *Journal of Geophysical Research, 91*(C6), 7691–7696.

Toyota, T., Takatsuji, S., & Nakayama, M. (2006). Characteristics of sea ice floe size distribution in the seasonal ice zone. *Geophysical Research Letters, 33*, L02616.

Tucker, W. B., & Perovich, D. K. (1992). Stress measurements in drifting pack ice. *Cold Regions Science and Technology, 20*(2), 119–139.

Turcotte, D. L. (1986). Fractals and fragmentation. *Journal of Geophysical Research, 91*(B2), 1921–1926.

Turcotte, D. L. (1992). *Fractals and chaos in geology and geophysics*. Cambridge: Cambridge University Press.

Utsu, T. (2002). Statistical features of seismicity. In Lee, W. H. K., Kanamori, H., Jennings, P. C., & Kisslinger, C. (Eds.), *International Handbook of Earthquake and Engineering Seismology* (pp. 719–732). New York: Academic Press.

van Doorn, E., Dhruva, B., & Sreenivasan, R. (2000). Statistics of wind direction and its increments. *Physics of Fluids, 12*(6), 1529–1534.

Weiss, J. (2003). Scaling of fracture and faulting in ice on Earth. *Surveys of Geophysics, 24*, 185–227.

Weiss, J. (2008). Intermittency of principal stress directions within Arctic Sea ice. *Physical Review E, 77*, 056106.

Weiss, J., & Marsan, D. (2004) Scale properties of sea ice deformation and fracturing. *Comptes Rendus Physique, 5*(7), 683–685.

Weiss, J., Marsan, D., & Rampal, P. (2009) Space and time scaling laws induced by the multiscale fracturing of the Arctic sea ice cover. In F. Borodich (Ed.), *IUTAM Symposium on scaling in solid mechanics*. (pp. 101–109), Springer.

Weiss, J., & Schulson, E. M. (2009). Coulombic faulting from the grain scale to the geophysical scale: Lessons from ice. *Journal of Physics. D. Applied Physics, 42*, 214017.

Weiss, J., Schulson, E. M., & Stern, H. L. (2007). Sea ice rheology from in situ, satellite and laboratory observations: Fracture and friction, *Earth Planet. Science Letters, 255*, 1–8.

Wells, D. L., & Coppersmith, K. J. (1994). New empirical relationships among magnitude, rupture length, rupture width, rupture area, and surface displacement. *Bulletin of the Seismological Society of America, 84*(4), 974–1002.

Chapter 5
Conclusion and Perspectives: Sea Ice Drift, Deformation and Fracturing in a Changing Arctic

Abstract This concluding chapter briefly reviews the recently evidenced modifications in Arctic sea ice drift, deformation and rheology that occurred during the last decades. For more than 30 years now, increasing average drift velocities and deformation rates, in both summer and winter, not only accompanied but most likely strengthened the Arctic sea ice decline through a modification of the albedo feedback. Indeed, a thinner ice cover means more fracturing, which in turns can have two consequences. First, more lead opening means a decreasing albedo. Second, increasing fracturing facilitates sea ice mobility and export out of the Arctic basin, i.e. means a negative contribution to mass balance. An average mechanical weakening of the Arctic sea ice cover has been highlighted from a strengthening of inertial oscillations. This interweaving of the sea ice state (thickness, concentration) in the one hand, and mechanical/dynamical processes on the other hand, calls for a continuing effort on the analysis of sea ice drift, deformation and fracturing.

This exploration of sea ice drift, deformation and fracturing has revealed the complexity of the processes involved, spanning a huge range of space and time scales, from the ice cover thickness (m) to the Arctic basin scale ($>10^6$ m), and from the timescale of icequakes (seconds) to seasons (months). This complexity arises from a combination of two factors that might be summarized in one phrase: A brittle elastic ice cover sandwiched between, and mechanically forced by, two turbulent fluids, the atmosphere and the ocean. Fluid turbulence is an emblematic example of a physical process involving features such as eddies interacting over a wide range of spatial and temporal scales. Consequently, following Thorndike's legacy, the statistical tools initially developed to analyze fluid turbulence have been applied to the study of sea ice kinematics. While similarities were noted on this basis, fundamental differences have been emphasized in Chaps. 2 and 3 intermittency and spatial heterogeneity are even more pronounced in the case of sea ice, and a detailed analysis of sea ice strain-rates revealed a space/time coupling whose equivalent has been observed, in terms of discrete events, for crustal deformation.

These properties of sea ice kinematics are likely the signature of the brittle nature of sea ice rheology, the second factor of complexity. Brittle rheology implies a strongly non-linear response of sea ice to mechanical forcing, with the consequence of reinforcing the intermittency and spatial localization, and modifying the scaling properties. This mechanical behavior has been explored and confirmed from an analysis of internal stresses and fracture networks in Chap. 4. As detailed below, it has also had important consequences for the evolution of the Arctic sea ice cover in the last few decades, and will have for the decades to come.

Indeed, so far we have considered sea ice kinematics and mechanics from a fundamental point of view, leaving aside the evolution of the Arctic ice cover in recent decades. There is now consensus towards a significant shrinking of the Arctic sea ice extent during the last few decades (Cavalieri and Parkinson 2012; Lemke et al. 2007; Serreze et al. 2007), with a spectacular acceleration of this decline within the last few years (Comiso et al. 2008; Serreze 2009; Stroeve et al. 2012a). The most emblematic symbol is the succession of record-breaking lows of the September Arctic sea ice minimum extent in 2002, 2005, 2007, and most recently in 2012 when the perennial ice cover shrank below 4 million km^2, compared to ~ 7.5 million km^2 in the early 1980's (Fetterer et al. 2002, updated 2012). This implies a spectacular shrinking of multiyear ice, especially in recent years (Kwok et al. 2009; Polyakov et al. 2012). In the meantime, the Arctic sea ice cover is thinning as well. Although there is not (yet) a long-term monitoring of ice thickness from satellite data at the basin scale, a combination of submarine ice draft measurements and satellite (ICESat) altimetry has revealed a net average thinning of the central Arctic ice pack from 1980 to 2008 of 1.75 m in winter, and 1.65 m in summer (Kwok and Rothrock 2009; Rothrock et al. 2008), representing thinning trends of respectively 17 % and 21 % per decade. The combination of decreasing extent and thinning implies a strongly negative Arctic sea ice mass balance over the last decades.

Note that, in contrast to the Arctic, the Antarctic sea ice cover has, on average, slightly expanded over the last decades, except around the Antarctic Peninsula (Parkinson and Cavalieri 2012). This might be due to changes in atmospheric circulation, or to increasing snow precipitations inducing increasing snow-ice formation (Powell et al. 2005).

This thinning and decline of Arctic sea ice has a strong impact on the positive ice-albedo feedback (e.g. Stroeve et al. 2012a; Winton 2008) over the Arctic Ocean during summer, which plays an essential role in the warming of northern latitudes (see e.g. Screen and Simmonds 2010; Screen et al. 2012; Serreze and Francis 2006 about Arctic amplification, and Fig. 5.1), and possibly of the entire planet. A strong impact on sea ice kinematics and mechanics is also expected. Firstly, the failure strength of the ice cover depends linearly on its thickness. Consequently, a thinner ice cover will break up and crumble more easily for the same wind/current mechanical forcing. In addition to this trivial effect of thickness, we can expect that a more fragmented, less concentrated sea ice cover will be more mobile. This can be qualitatively understood from the classical force balance (Fig. 1.2): in a looser ice pack, the "internal friction" term $\nabla \cdot \sigma$ of the momentum balance (Eq. 1.1) is expected to decrease and therefore to offer less counterbalance to the effect of wind/current forcing on sea ice motion. In the case of vanishing internal friction, free drift is obtained as a limiting case (see the introduction of Chap. 2).

5 Conclusion and Perspectives: Sea Ice Drift, Deformation

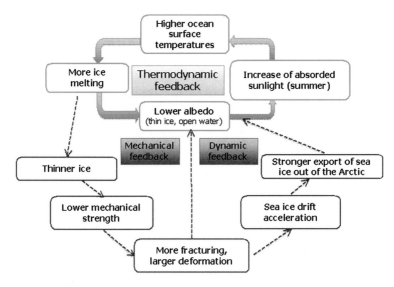

Fig. 5.1 The albedo feedback for an ice-covered ocean. The upper loop (*thick arrows*) involves only radiative and thermodynamic processes. This albedo feedback is strengthened by mechanical and dynamical feedbacks (see text for details)

Furthermore, an increase of sea ice fracturing and mobility will likely reinforce its decline through two main mechanisms (Fig. 5.1): First, during summer, increasing fracturing means more lead opening and therefore a decreasing albedo (Zhang et al. 2000). A warming of the Arctic ocean, in turn, enhances sea ice thinning in summer and delays refreezing in early winter. Second, and independently of the season, increasing mobility accelerates the export of sea ice through Fram Strait, with a possibly significant impact on sea ice mass balance (Haas et al. 2008; Rampal et al. 2009a; Rampal 2011). Similarly, sea ice mechanical weakening decreases the likelihood of arch formation across Nares Strait, therefore allowing the old, thick ice that covers this part of the Arctic Ocean to be exported through this strait (Kwok et al. 2010). Therefore, several questions arise: (1) Did the mobility of Arctic sea ice increase in recent decades?, (2) Do we have observational evidence for a recent mechanical weakening of the ice cover which could explain this accelerated kinematics?, and (3) What could be the impact of this on Arctic sea ice mass balance and evolution?

In 2006–2007, the polar schooner *Tara* repeated the *Fram*'s journey along the transpolar drift (Fig. 5.2): she became locked in the ice in the Laptev Sea in early September 2006, but crossed Fram Strait only 16 months later in December 2007, drifting on average more than twice as fast as the *Fram* 115 years before (Gascard 2008). Was this surprisingly fast journey just by chance, due to a specific Arctic atmospheric circulation pattern enhancing the transpolar drift and therefore sea ice export through Fram Strait (Lindsay et al. 2009; Wang et al. 2009)? Or was it a sign of a more fundamental, long-term evolution of sea ice kinematics?

Fig. 5.2 The polar schooner *Tara*, locked in the ice, during its journey along the Transpolar Drift in April 2007 (photo J.P. Metaxian)

To explore this, Rampal et al. (2009a) analyzed the evolution of Arctic sea ice drift velocities and strain-rates from 1979 to 2007 from the IABP dataset described in Sect. 2.1. To estimate sea ice deformation rates, they used the strain-rate proxy $\dot{\varepsilon}_{disp}$ based on the dispersion of pairs of tracers (Eq. 3.5). They reported that the sea ice mean speed over the Arctic increased at a rate of 17.0($\pm$4.5) % per decade in winter and 8.5($\pm$2.0) % per decade in summer from 1979 to 2007 (Fig. 5.3), whereas the mean deformation rate increased more than 50($\pm$10) % per decade in both summer and winter over the same period (Fig. 5.4). This spectacular evolution of sea ice kinematics was recently confirmed over a shorter period (1992–2009) from sea ice drift speeds derived from SSM/I satellite data, with an average trend of 10.6($\pm$0.9) %/decade (Spreen et al. 2011). The question that arises is whether this acceleration results from a concomitant increase of wind forcing, or results from a modification of the sea ice state (thickness, concentration and mechanical strength).

Analyzing the average 10 m-height wind speed over the Arctic from the ERA-40 reanalysis dataset, Rampal et al. (2009a) obtained an insignificant trend of +0.4 %/decade and therefore attributed most of the acceleration of sea ice drift and deformation to a modification of the sea ice state. Examining the correspondence between wind and sea ice speed trends over different regions of the Arctic, Spreen et al. (2011) concluded that a fraction of the observed increase in sea ice speed could be explained by a change in wind speed in the central Arctic, but not elsewhere. Considering correlations between the average sea ice speed and atmospheric circulation indices, Vihma et al. (2012) concluded that while interannual variations of Arctic sea ice speed can be partly explained by atmospheric forcing, the long-term positive trend

5 Conclusion and Perspectives: Sea Ice Drift, Deformation

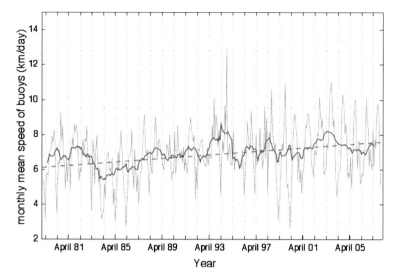

Fig. 5.3 *Thin solid line*: monthly mean sea ice speed within the Arctic Ocean, from January 1979 to December 2007, obtained from the IABP dataset. *Thick solid line*: 12-month running mean. A linear fit of the data is plotted as a *dashed line*, and gives a slope of 0.56(±0.11) km/day per decade (adapted from Rampal et al. 2009a)

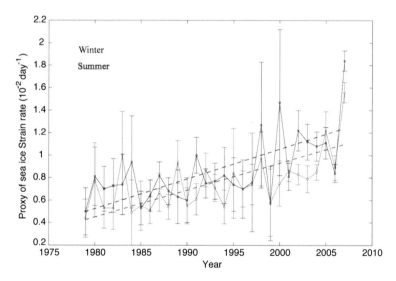

Fig. 5.4 Seasonal average sea ice strain-rate proxy in the Arctic basin from 1979 to 2007. The proxy $\dot{\varepsilon}_{disp}$ is calculated from the dispersion rate of pairs of tracers for timescales shorter than 1 day and spatial scales ranging from 50 to 500 km (see Rampal et al. 2009a). *Error bars* are calculated from a bootstrap method, and the weighted linear fits of the data, plotted as *dashed lines*, give trends of 2.4(±0.4) × 10^{-4} day^{-1} per year for winter (i.e. 51 ± 8.5 % increase per decade), and 2.6(±0.6) × 10^{-4} day^{-1} per year for summer (i.e. 52 ± 12 % increase per decade). Adapted from Rampal et al. (2009a)

cannot. Therefore, the observed acceleration of sea ice kinematics over the last few decades does not result from increasing external forcing, but instead from a modification of the sea ice state.

At this stage, the question remains whether the observed sea ice decline (thinning and decreasing extent) could account for this accelerated kinematics through a "trivial" effect on the sea ice momentum balance through a decreasing ice mass (see Eq. 1.1), or if a more subtle evolution of the sea ice internal friction could play a role through a mechanical weakening. To estimate sea ice strength at the basin scale is a difficult task: the internal stress measurements discussed in Sect. 4.2 are too limited in terms of spatial and temporal coverage. Gimbert et al. (2012a, b) proposed an analysis at the basin and multidecadal scales based on a specific "rheometer", the response of sea ice to the well-defined Coriolis force. As this specific forcing is constant over time, an evolution of ice motion around the inertial frequency (≈ 2 cycles/day in the Arctic) would be a signature of a modification of sea ice mechanical behaviour. From the IABP dataset, Gimbert et al. (2012a) revealed a significant strengthening of sea ice inertial motion at the basin scale, in both winter and summer, in recent years (Fig. 5.5). Once again, it is necessary to differentiate the effect of ice thinning (and possibly of decreasing concentration) from that of an actual mechanical weakening. Gimbert et al.(2012b) presented a simple ice-ocean boundary layer coupled model in which sea ice rheology and internal friction are simplified through a linear term, $-KU_i$, where K is a large scale (apparent) friction coefficient that represents the inverse of a temporal scale of dissipation. This linear friction term is extremely crude with regard to the elasto-brittle, strongly non-linear behaviour discussed in Chap. 4. However, with proper averaging over large spatial (almost the basin scale) and temporal (seasonal) scales, such a model describes well the sea ice behaviour in the frequency domain, and allows us to differentiate the effect of ice concentration, thickness and internal friction on sea ice inertial motion. Gimbert et al. (2012b) revealed in this way a strong decrease of the internal friction coefficient K, i.e. a genuine mechanical weakening of the sea ice cover in recent years (2002–2008) compared to the former period (1979–2001).

Analyses of ice drifter trajectories thus answered positively the two questions raised above, validating the scenario of a strengthening of the classical ice-albedo feedback loop through mechanical/dynamical processes (Fig. 5.1).

To conclude this volume on sea ice drift, deformation and fracturing, and considering their role in the ice-albedo feedback and negative mass balance described above, once can raise the question of the potential role of these processes for the future of the Arctic sea ice cover. The usual way to forecast this evolution is to run climate simulations. Owing to the role of the ice cover on the global climate, all current climate models include a sea ice model which, in most cases, couples a thermodynamic component to a dynamical/mechanical component. However, climate models, including state-of-the-art ones, underestimate the observed sea ice decline either in terms of sea ice extent (Stroeve et al. 2007, 2012b), or thickness (Rampal et al. 2011), and did not foresee its recent acceleration (Serreze et al. 2009). This therefore raises uncertainties in the Arctic sea ice cover and climate projections for the twenty first century. Several explanations have been proposed to

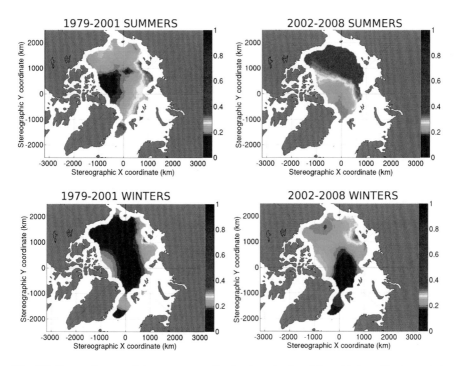

Fig. 5.5 Seasonal maps of the average amplitude of sea ice inertial oscillations, for the period 1979–2001 (*left*) and the period 2002–2008 (*right*), as measured by a dimensionless parameter which is close to 1 in case of free drift, and equal to 0 in the absence of detectable inertial oscillations. See Gimbert et al. (2012a) for details. This shows that sea ice behaved nearly in free drift in recent years during summer in the Beaufort Sea as well as in the eastern Arctic, and that inertial oscillations became more pronounced even in winter, thus arguing for a more mobile, weaker sea ice cover

explain these deficiencies, including incorrect representations of the albedo feedback (Winton 2008), of the formation of melt ponds (Flocco et al. 2010; Skyllingstad et al. 2009), of the role of polar clouds (Eisenman et al. 2007; Vavrus et al. 2009), and of the modes of atmospheric variability that can impact the sea ice general circulation and so the ice export through Fram Strait (Kwok et al. 2004; Nghiem et al. 2007; Smedsrud et al. 2011). Increasing heat fluxes from Pacific Waters through Bering Strait (Woodgate et al. 2010), or from Atlantic Waters from Fram Strait (Polyakov et al. 2010), may also have play a role on the recent sea ice decline, and the ability of climate models to correctly simulate these processes should be checked.

Rampal et al. (2011) argued that an inaccurate representation of sea ice mechanics and kinematics may also be an important shortcoming, as it would not allow the correct representation of the strengthening of the sea ice decline through mechanical processes (see Fig. 5.1). They showed that CMIP3 climate models, developed in the context of the IPCC Fourth Assessment Report, all failed to capture the acceleration of ice motion detailed above (Fig. 5.3), as the result of an unexpectedly weak coupling between the ice state (thickness, concentration) and the ice velocity. In other words, sea ice behaves in these simulations as drifting almost freely. This is surprising, owing to the large range of complexity of the associated sea ice models, from a sea ice cover modeled as a perfect fluid (without internal stress), to a viscous-plastic rheology (VP, Hibler 1979) combined with a redistribution scheme of ice thickness through deformation processes (Rothrock 1975).

Most state-of-the-art models that incorporate a sea ice rheology use the classical VP scheme in which sea ice has strength under convergence and shear, but not under divergence. For stress states inside the plastic yield curve, sea ice behaves as a viscous fluid, while it flows as a perfect plastic once the stress state reaches this yield curve [see e.g. Feltham (2008) for a review of sea ice rheological models]. This rheology is in contradiction with the analysis of sea ice internal stresses that revealed a Coulombic brittle rheology (see Sect. 4.2 and Weiss et al. 2007). However, the key point to raise here is that with such a VP framework, when an ice parcel locally fails (i.e. plastically flows), the elastic stresses are not redistributed, i.e. long-range interactions are absent. Consequently, such models are unable to reproduce the scaling laws of sea ice deformation detailed in Sect. 3.2, which are the signature of these long-range elastic interactions (Girard et al. 2009). These shortcomings likely imply an underestimation of the non-linear relationship between the sea ice state and sea ice strength, i.e. of the effect of sea ice mechanics on the albedo feedback (Fig. 5.1). While this might partly explain why climate models underestimate the Arctic sea ice decline, the free drift-like behavior of modeled sea ice remains surprising (Rampal et al. 2011). Indeed, in the VP mechanical model, a coupling between the ice state and kinematics is expected to take place through the acceleration term of the momentum equation (1.1) that depends on the ice mass per unit area, and through the ice strength P that defines the size of the plastic envelope (Hibler 1979). One may ask whether a modified parameterization of the VP scheme might improve this point, or if the development of entirely new frameworks taking into account the elasto-brittle, frictional character of sea ice mechanics are necessary (e.g. Girard et al. 2011; Hopkins et al. 2004; Wilchinsky et al. 2010).

Either way, Rampal et al. (2011) showed that the models' deficiencies in reproducing the recent kinematic evolution of the ice cover may have a strong impact on the simulated sea ice mass balance in the Arctic, as they imply an underestimation of sea ice export. Imposing a positive trend of 12 %/decade on modeled sea ice speeds at the main ice flux gates surrounding the Arctic Basin, (a value close to the observed drift acceleration within this basin (Rampal et al. 2009a; Spreen et al. 2011)), they were able to fully reduce the mismatch between modeled and observed sea ice area in September (Fig. 5.6). This means that the projections for an ice-free summer in the Arctic by 2100 are most likely too conservative, and that such a situation might happen within a few decades.

This discussion on the future of the sea ice cover, with strong implications for the Earth's climate, calls for a concerted effort on the fundamental analysis of sea ice drift, deformation and fracturing, and on the ways to properly model these

5 Conclusion and Perspectives: Sea Ice Drift, Deformation

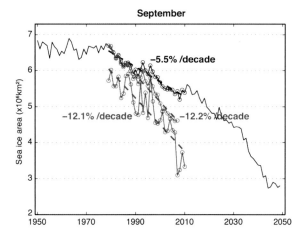

Fig. 5.6 September Arctic sea ice area. The *black line* shows the ensemble mean of the IPCC AR4 simulations, whereas satellite-derived estimates are plotted as *red lines + circles*. Corresponding linear fits over the period 1979–2010 are shown as *thick dashed lines*. The mismatch between modeled and observed trends can be removed by imposing a positive trend of 12 %/decade on the modeled ice speeds at certain flux gates over the period 1979–2007 (*blue lines*) (from Rampal et al. 2011)

processes. Initiated 120 years ago by Nansen's 2 % rule of thumb (Nansen 1902) (see Chap. 1), developed by Thorndike and co-workers in the 1980s, and reopened in recent years, this research topic is certainly still ripe for future work.

References

Cavalieri, D. J., & Parkinson, C. L. (2012). Arctic sea ice variability and trends, 1979–2010. *Cryosphere, 6*(4), 881–889.

Comiso, J. C., Parkinson, C. L., Gersten, R., & Stock, L. (2008). Accelerated decline in the Arctic sea ice cover. *Geophysical Research Letters, 35*, L01703.

Eisenman, I., Untersteiner, N., & Wettlaufer, J. S. (2007). On the reliability of simulated Arctic sea ice in global climate models. *Geophysical Research Letters, 34*(10), L10501.

Fetterer, F., Knowles, K., Meier, W., & Savoie, M. (2002). *updated 2012)*. National Snow and Ice Data Center, Boulder, Colorado, USA: Sea ice index.

Feltham, D. (2008). Sea ice rheology. *Annual Review of Fluid Mechanics, 40*, 91–112.

Flocco, D., Feltham, D.L. & Turner A.K. (2010). Incorporation of a physically based melt pond scheme into the sea ice component of a climate model, *Journal of Geophysical Research-Oceans, 115*, C08012.

Gascard, J. C., et al. (2008). Exploring Arctic transpolar drift during dramatic sea ice retreat. *EOS, 89*(3), 21–22.

Gimbert, F., N. C. Jourdain, D. Marsan, J. Weiss, and B. Barnier (2012b), Recent mechanical weakening of the Arctic sea ice cover as revealed from larger inertial oscillations, *Journal of Geophysical Research, 117*, C00J12.

Gimbert, F., Marsan, D., Weiss, J., Jourdain, N. C., & Barnier, B. (2012a). Sea ice inertial oscillation magnitudes in the Arctic basin. *The Cryosphere, 6*, 1187–1201.

Girard, L., Weiss, J., Molines, J. M., Barnier, B., & Boullion, S. (2009). Evaluation of two sea ice models on the basis of statistical and scaling properties of Arctic sea ice deformation. *Journal of Geophysical Research, 114*, C08015.

Girard, L., Bouillon, S., Weiss, J., Amitrano, D., Fichefet, T., & Legat, V. (2011). A new modelling framework for sea-ice mechanics based on elasto-brittle rheology. *Annals of Glaciology, 52*(57), 123–132.

Haas, C., Pfaffling, A., Hendricks, S., Rabenstein, L., Etienne, J. L., & Rigor, I. (2008). Reduced ice thickness in Arctic transpolar drift favors rapid ice retreat. *Geophysical Research Letters, 35*, L17501.

Hibler, W. D. I. (1979). A dynamic thermodynamics sea ice model. *Journal of Physical Oceanography, 9*, 815–846.

Hopkins, M. A., Frankenstein, S., & Thorndike, A. S. (2004). Formation of an aggregate scale in Arctic sea ice. *Journal of Geophysical Research, 109*, C01032.

Kwok, R., & Rothrock, D. A. (2009). Decline in Arctic sea ice thickness from submarine and ICES at records: 1958–2008. *Geophysical Research Letters, 36*, L15501.

Kwok, R., Pedersen, L. T., Gudmandsen, P., & Pang, S. S. (2010). Large sea ice outflow into the Nares Strait in 2007. *Geophysical Research Letters, 37*, L03502.

Kwok, R., Cunningham, G. F., & Pang, S. S. (2004). Fram Strait sea ice outflow. *Journal of Geophysical Research, 109*, C01009.

Kwok, R., Cunningham, G. F., Wensnahan, M., Rigor I., H. J. Zwally & Yi D. (2009). Thinning and volume loss of the Arctic Ocean sea ice cover: 2003–2008, *Journal of Geophysical Research Oceans, 114*.

Lemke, P., et al. (2007). Observations: Changes in snow, ice and frozen ground, in *Climate Change 2007: The physical basis. Contribution of working group I to the fourth assessment report of the Intergovernmental Panel on Climate Change*, (S. Solomon, D. Qin, M. Manning, Z. Chen, M. Marquis, M. Averyt, M. Tignor and H. L. Miller eds.), Cambridge: Cambridge University Press.

Lindsay, R. W., Zhang, J., Schweiger, A., Steele, M., & Stern, H. (2009). Arctic sea ice retreat in 2007 follows thinning trend, J. *Journal of Climate, 22*, 165–175.

Nansen, F. (1902). Oceanography of the north polar basin: The Norwegian north polar expedition 1893–1896, Scientific Results, 3(9).

Nghiem, S. V., Rigor, I. G., Perovich, D. K., Clemente-Colon, P., Weatherly, J. W., & Neumann, G. (2007). Rapid reduction of Arctic perennial sea ice. *Geophysical Research Letters, 34*, L19504.

Parkinson, C. L., & Cavalieri, D. J. (2012). Antarctic sea ice variability and trends, 1979–2010. *The Cryosphere, 6*(4), 871–880.

Polyakov, I. V., Walsh, J. E., & Kwok, R. (2012). Recent Changes of Arctic Multiyear Sea Ice Coverage and the Likely Causes. *Bulletin of the American Meteorological Society, 93*(2), 145–151.

Polyakov, I. V., et al. (2010). Arctic Ocean Warming Contributes to Reduced Polar Ice Cap. *Journal of Physical Oceanography, 40*(12), 2743–2756.

Powell, D. C., T. Markus, & A. Stossel. (2005). Effects of snow depth forcing on Southern Ocean sea ice simulations, *Journal of Geophysical Research-Oceans, 110*(C6).

Rampal, P., Weiss, J., & Marsan, D. (2009). Positive trend in the mean speed and deformation rate of Arctic sea ice: 1979–2007. *Journal of Geophysical Research, 114*, C05013.

Rampal, P., J. Weiss, C. Dubois, & J. M. Campin. (2011). IPCC climate models do not capture Arctic sea ice drift acceleration: Consequences in terms of projected sea ice thinning and decline, *Journal of Geophysical Research, 116*, C00D07.

Rothrock, D. A., Percival, D. B., & Wensnahan, M. (2008). The decline in arctic sea-ice thickness: Separating the spatial, annual, and interannual variability in a quarter century of submarine data. *Journal of Geophysical Research, 113*, C05003.

Rothrock, D. A. (1975). The energetics of the plastic deformation of pack ice by ridging. *Journal of Geophysical Research, 80*(33), 4514–4519.

References

Serreze, M. C., Holland, M. M., & Stroeve, J. (2007). Perspectives on the Arctic's shrinking sea-ice cover. *Science, 315*, 1533–1536.

Serreze, M. C. (2009). Arctic climate change: Where reality exceeds expectations. *Arctic, 13*(1), 1–4.

Smedsrud, L. H., Sirevaag, S., Kloster, K., Sorteberg, A., & Sandven, S. (2011). Recent wind driven high sea ice area export in the Fram Strait contributes to Arctic sea ice decline. *The Cryosphere, 5*, 821–829.

Screen, J. A., & Simmonds, I. (2010). The central role of diminishing sea ice in recent Arctic temperature amplification. *Nature, 464*, 1334–1337.

Screen, J. A., Deser, C., & Simmonds, I. (2012). Local and remote controls on observed Arctic warming. *Geophysical Research Letters, 39*, L10709.

Serreze, M. C., & Francis, J. A. (2006). The arctic amplification debate. *Climate Change, 76*(3–4), 241–264.

Spreen, G., Kwok, R., & Menemenlis, D. (2011). Trends in Arctic sea ice drift and role of wind forcing: 1992–2009. *Geophysical Research Letters, 38*, L19501.

Stroeve, J., Holland, M. M., Meier, W., Scambos, T., & Serreze, M. (2007). Arctic sea ice decline: Faster than forecast. *Geophysical Research Letters, 34*, L09501.

Stroeve, J. C., Serreze, M. C., Holland, M. M., Kay, J. E., Malanik, J., & Barrett, A. P. (2012a). The Arctic's rapidly shrinking sea ice cover: a research synthesis. *Climate Change, 110*(3–4), 1005–1027.

Stroeve, J. C., Kattsov, V., Barrett, A., Serreze, M., Pavlova, T., Holland, M., et al. (2012b). Trends in Arctic sea ice extent from CMIP5, CMIP3 and observations. *Geophysical Research Letters, 39*, L16502.

Skyllingstad, E. D., Paulson C. A., & Perovich D. K. (2009). Simulation of melt pond evolution on level ice, *Journal of Geophysical Research-Oceans, 114*, C12019.

Vavrus, S., Waliser, D., Schweiger, A., & Francis, J. (2009). Simulations of 20th and 21st century Arctic cloud amount in the global climate models assessed in the IPCC AR4. *Climate Dynamics, 33*(7–8), 1099–1115.

Vihma, T., Tisler, P., & Uotila, P. (2012). Atmospheric forcing on the drift of Arctic sea ice in 1989–2009. *Geophysical Research Letters, 39*, L02501.

Wang, J., Zhang, J., Watanabe, E., Ikeda, M., Mizobata, K., Walsh, J. E., et al. (2009). Is the dipole anomaly a major driver to record lows in Arctic summer sea ice extent? *Geophysical Research Letters, 36*, L05706.

Weiss, J., Schulson, E. M., & Stern, H. L. (2007). Sea ice rheology from in situ, satellite and laboratory observations: Fracture and friction, *Earth Planet. Earth and Planetary Science Letters, 255*, 1–8.

Wilchinsky, A. V., D. L. Feltham, and M. A. Hopkins (2010), Effect of shear rupture on aggregate scale formation in sea ice, *Journal of Geophysical Research-Oceans, 115*.

Winton, M. (2008). Sea ice-albedo feedback and Nonlinear arctic climate change. In E. T. DeWeaver, C. M. Bitz, & L. B. Tremblay (Eds.), *Arctic Sea Ice Decline: Observations, Projections, Mechanisms, and Implications* (pp. 111–131). Washington: American Geophysical Union.

Woodgate, R. A., Weingartner, T., & Lindsay, R. (2010). The 2007 Bering Strait oceanic heat flux and anomalous Arctic sea-ice retreat. *Geophysical Research Letters, 37*, L01602.

Zhang, J., Rothrock, D., & Steele, M. (2000). Recent changes in arctic sea ice: the interplay between ice dynamics and thermodynamics. *Journal of Climate, 13*(17), 3099–3114.

DISCARDED
CONCORDIA UNIV. LIBRARY

CONCORDIA UNIVERSITY LIBRARIES
MONTREAL

Printed by Publishers' Graphics LLC
LMO130508.15.16.46